MEP 701A 3KW Generator Pair
Operators and Maintenance Manual
TM 5-6115-640-14&P
Includes Repair Parts and Special Tools List

For Power Plants:
AN/MJQ-32 and AN/MJQ-33

On 2 Wheel 3/4 Ton M116A2 Trailer

Generator Set Pair
Trailer Mounted, Tactical Quiet

edited by
Brian Greul

The MEP series of Military Generators are rugged, durable and incorporate proven diesel engine technology. This book is the the combined Operator, Maintenance, and Parts List Manual. It is being republished to assist enthusiasts, restorers, and aftermarket owners who use or wish to use these generators outside of military use.

An 8.5x11 3 hole punched loose leaf copy may be purchased for your 3 ring binder. Email books@ocotillopress.com for current information.

Should you have suggestions or feedback on ways to improve this book please send email to Books@OcotilloPress.com

Edited 2021 Ocotillo Press
ISBN 978-1-954285-24-8

Printed in the United States of America

Ocotillo Press
Houston, TX 77017
Books@OcotilloPress.com

TECHNICAL MANUAL

OPERATOR, UNIT, DIRECT SUPPORT AND GENERAL
SUPPORT MAINTENANCE MANUAL (INCLUDING
REPAIR PARTS AND SPECIAL TOOLS LIST)

POWER PLANTS

AN/MJQ-32 (NSN 6115-01-280-2300)

AN/MJQ-33 (NSN 6115-01-280-2301)

(2ea.) MEP-701A 3 KW 60 HZ

ACOUSTIC SUPPRESSION KIT GENERATOR SETS

M116A2 2-WHEEL, 2-TIRE, 3/4-TON

MODIFIED TRAILERS

Approved for public release; Distribution is unlimited.

This copy is a reprint which includes current

HEADQUARTERS, DEPARTMENT OF THE ARMY
28 JULY 1989

CHANGE HEADQUARTERS

DEPARTMENT OF THE ARMY

NO. 2 WASHINGTON, D.C., 17 September 1991

Operator, Unit, Direct Support and General Support
Maintenance Manual (Including Repair Parts and Special Tools List)

**POWER PLANTS
AN/MJQ-32 (NSN 6115-01-280-2300)
AN/MJQ-33 (NSN 6115-01-280-2301)
(2 ea.) MEP-701A 3 KW 60 HZ
ACOUSTIC SUPPRESSION KIT GENERATOR SETS
M116A2 2-WHEEL, 2-TIRE, 3/4-TON
MODIFIED TRAILERS**

Approved for public release; distribution is unlimited TM

5-6115-640-14&P, 28 July 1989, is changed as follows:

1. Remove and insert pages as indicated below. New or changed text material is indicated by a vertical bar in the margin. An illustration change is indicated by a miniature pointing hand.

Remove pages Insert pages

1-5 through 1-8 1-5 through 1-8
4-23 through 4-26 4-23 through 4-26
C-53 through C-56 C-53 through C-56
D-7 and D-8 D-7 and D-8

2. Retain this sheet in front of manual for reference purposes.

By Order of the Secretary of the Army:

GORDON R. SULLIVAN
General, United States Army
Chief of Staff

Official:
PATRICIA P. HICKERSON
Brigadier General, United States Army
The Adjutant General

DISTRIBUTION:
To be distributed in accordance with DA Form 12-25E, (qty rqr block No. 4197)

CHANGE HEADQUARTERS

NO. 1 WASHINGTON, DC, 20 February 1990

DEPARTMENT OF THE ARMY

Operator, Unit, Direct Support and General
Support Maintenance Manual (Including
Repair Parts and Special Tools List)

**POWER PLANTS
AN/MJQ-32 (NSN 6115-01-280-2300)
AN/MJQ-33 (NSN 6115-01-280-2301)
(2ea.) MEP-701A 3 KW 60 HZ
ACOUSTIC SUPPRESSION KIT GENERATOR SETS
M116A2 2-WHEEL, 2-TIRE, 3/4-TON
MODIFIED TRAILERS**

Approved for public release; distribution is unlimited TM

5-6115-640-14&P, 28 July 1989, is changed as follows:

1. Remove and insert pages as indicated below. New or changed text material is indicated by a vertical bar in the margin. An illustration change is indicated by a miniature pointing hand.

Remove pages Insert pages

e and f e and f A-1 and A-2 A-1 and A-2
D-7 and D-8 D-7 and D-8

2. Retain this sheet in front of manual for reference purposes.

By Order of Secretary of the Army:

CARL E. VUONO
General, United States Army
Chief of Staff

Official:
WILLIAM J. MEEHAN, II
Brigadier General, United States Army
The Adjutant General

DISTRIBUTION:
    To be distributed in accordance with DA Form 12-25A, Operator, Unit, Direct Support and General Support Maintenance requirements for Generator Set, Diesel Engines Driven, Trailer Mounted

5 SAFETY STEPS TO FOLLOW IF SOMEONE IS THE VICTIM OF ELECTRICAL SHOCK

1 DO NOT TRY TO PULL OR GRAB THE INDIVIDUAL

2 IF POSSIBLE, TURN OFF THE ELECTRICAL POWER

3 IF YOU CANNOT TURN OFF THE ELECTRICAL POWER, PULL, PUSH, OR LIFT THE PERSON TO SAFETY USING A WOODEN POLE OR A ROPE OR SOME OTHER INSULATING MATERIAL

4 SEND FOR HELP AS SOON AS POSSIBLE

5 AFTER THE INJURED PERSON IS FREE OF CONTACT WITH THE SOURCE OF ELECTRICAL SHOCK, MOVE THE PERSON A SHORT DISTANCE AWAY AND IMMEDIATELY START ARTIFICIAL RESUSCITATION

WARNING
All specific cautions and warnings contained in this manual shall be strictly adhered to. Otherwise, severe injury or death to personnel or damage to the equipment may result.

WARNING
Do not operate generator sets until power plant is properly grounded. Serious injury or death by electrocution can result from operating an ungrounded power plant.

WARNING
Clean parts in a well-ventilated area. Avoid inhalation of solvent fumes and prolonged exposure of skin to cleaning solvent. Wash exposed skin thoroughly. Dry cleaning solvent (PD-680) used to clean parts is potentially dangerous to personnel and property. Do not smoke or use near open flame or excessive heat. Flash point of solvent is 1000F to 138°F (38°C to 59°C).

WARNING
Before performing any maintenance that requires climbing on or under trailer, set trailer hand-brakes, chock wheels, and lower rear leg prop. Injury to personnel could result from trailer suddenly rolling or tipping.

WARNING

Steel strapping used in packaging of the power plants has sharp edges. Care should be taken when cutting and handling strapping to avoid injury to personnel.

WARNING

Basic Issue Items List (BIIL) box weighs approximately 225 lb. Use at least two men when removing box from stowage rack to avoid injury to personnel.

WARNING

Bow assembly will fan and cause injury to personnel if not supported before cutting away steel strappings.

WARNING

Remove fire extinguishers and fuel cans prior to start-up of generator. This will ensure that in the event of fire extra fuel will not be involved and extinguisher will remain accessible.
WARN ING Make sure generator sets are shut down before performing any continuity checks on switch box. Failure to do so may result in injury or death by electrocution.

c

WARNING

Make sure generator set circuit breakers are in the OFF position before performing removal procedures on switch box. Failure to follow this precaution may result in injury or death by electrocution.

WARNING

Make sure generator sets are shut down before performing any maintenance on wires or cables. Failure to follow this precaution may result in injury or death by electrocution.

WARNING

When lifting generator set, use lifting equipment with a minimum lifting capacity of 1500 lb. Do not stand under generator while it is being lifted. Failure to observe these precautions can cause death or injury to personnel or damage to equipment.

WARNING

When lifting trailer body, use lifting equipment with a minimum lifting capacity of 500 lb. Do not stand under trailer body while it is being lifted. Failure to observe these precautions can cause death or injury to personnel or damage to equipment.

d

WARNING

When angled bar is removed, rear leg prop assembly on trailer chassis will fall from bracket if not supported. To prevent injury to personnel or damage to equipment, do not permit rear leg prop assembly to drop.

WARNING

When operating the Power Plant off road, the operator must conduct a visual inspection of the trailer shock absorbers and trailer brake actuator at each stop. This must not exceed 50 miles of operation.

WARNING

While in transit, ensure that the trailer legs are fully retracted, and that the retainer pins are properly inserted, and in a downward position.

WARNING

While in transit, operators must perform visual checks of the trailer support legs at each stop to ensure that the pins are securely in place.

WARNING

When operating a Power Plant in temperature conditions exceeding 120 degrees fahrenheit (49 degrees centigrade), the operator must be aware of and perform checks of engine oil levels and oil viscosity more frequently than shown in the Preventive Maintenance Checks and Service table. Interval shown in this table is for normal operating temperatures.

WARNING

When performing operational tasks, i.e., cleaning, camouflage net installation, etc., operators are warned that non-skid surfaces have not been added to this configuration. Extreme caution must be used when climbing on the Power Plants. Under no condition should an operator stand on top of the generator or the acoustical suppression kit.

WARNING

The fitted cover provided with the AN/MJQ-33, Power Plant, is not to be used as a personnel shelter. This fitted cover must be rolled up when generator engine is running, otherwise asphyxiation of personnel could occur.

WARNING

To avoid asphyxiation do not operate the Power Plant within a 25 foot radius of any structure.

WARNING

Never operate the Power Plant in an enclosed area. Maintenance can be performed in a closed area providing and exhaust duct is used an adequate ventilation exists.

WARNING

Do not add fuel in the onboard fuel tank of an operating generator set.

WARNING

To avoid serious burns let the generator set cool before performing operator/maintenance checks and services.

WARNING

Information regarding NBC decontamination of this equipment is contained in FM 3-5.

change 1   f

TECHNICAL MANUAL HEADQUARTERS

DEPARTMENT OF THE ARMY

No. 5-6115-640-14&P WASHINGTON, DC, 28 July 1989

Operator, Unit, Direct Support and General Support Maintenance Manual (Including Repair Parts and Special Tools Lists) for POWER PLANTS AN/MJQ-32 (NSN 6115-01-280-2300) AND AN/MJQ-33 (NSN 6115-01-28U-2301) (2 EA) MEP-701A 3 KW 60 HZ ACOUSTIC SUPPRESSION KIT GENERATOR SETS M116A2 2-WHEEL, 2-TIRE, 3/4-TON MODIFIED TRAILERS

REPORTING ERRORS AND RECOMMENDING IMPROVEMENTS

You can help improve this manual. If you find any mistakes, or if you know of a way to improve the procedures, please let us know. Mail your letter, DA Form 2028 (Recommended Changes to Publications and Blank Forms), or DA Form 2028-2 located in the back of this manual direct to: Commander, U.S. Army Troop Support Command, ATTN: AMSTRMCTS, 4300 Goodfellow Boulevard, St. Louis, MO 63120-1798.

In either case, a reply will be furnished to you.

Approved for public release; distribution is unlimited.

Page

CHAPTER 4 UNIT MAINTENANCE

CHAPTER 5 DIRECT SUPPORT AND GENERAL SUPPORT MAINTENANCE INSTRUCTIONS

CHAPTER 6 TEST AND INSPECTION AFTER REPAIR

iii

LIST OF ILLUSTRATIONS

LIST OF ILLUSTRATIONS - Continued

LIST OF TABLES

v/(vi blank)

CHAPTER 1
INTRODUCTION

Section I. GENERAL

1-1. SCOPE.

This manual is for your use in operating and maintaining the Power Plants, AN/MJQ-32 and AN/MJQ-33. Both power plants are mobile units used to supply power to any system or equipment requiring up to 3 kW of 60 Hz input operating power. In addition to operating instructions and operator, unit, direct support, and general support maintenance procedures, this manual contains a Repair Parts and Special Tools List (RPSTL) for the power plants.

1-2. MAINTENANCE FORMS AND RECORDS.

Maintenance Forms and Records are prescribed by DA Pam 738-750.

1-3. REPORTING OF ERRORS.

Reporting of errors and omissions and recommendations for improvement of this publication by the individual user is encouraged. Reports should be submitted as follows: Mail your letter, DA Form 2028 (Recommended Changes to Publications and Blank Forms), or DA Form 2028-2 located in the back of this manual direct to Commander, U.S. Army Troop Support Command, ATTN: AMSTR-MCTS, 4300 Goodfellow Boulevard, St. Louis, MO 63120-1798.

1-4. REPORTING EQUIPMENT IMPROVEMENT RECOMMENDATIONS (EIR).

EIR's can and must be submitted by anyone who is aware of an unsatisfactory condition with the equipment design or use. It is not necessary to show a new design or list a better way to perform a procedure. Just simply tell why the design is unfavorable or why the procedure is difficult. EIR's may be submitted on SF 368. Mail directly to Commander, U.S. Army Troop Support Command, ATTN: AMSTR-QX, 4300 Goodfellow Boulevard, St. Louis, MO 63120-1798. A reply will be furnished to you.

1-5. LEVEL OF MAINTENANCE ACCOMPLISHMENT.

Refer to the Maintenance Allocation Chart (MAC) for tasks and levels of maintenance to be performed. **1-6. DESTRUCTION OF ARMY MATERIEL.**

Destruction of Army materiel to prevent enemy use shall be in accordance with TM 750-244-3.

Section II. EQUIPMENT DESCRIPTION AND DATA

1-7. DESCRIPTION.

Power Plants (PP's) AN/KJQ-32 and AN/KIQ-33 are each made up of two Tactical Utility Acoustic Suppression Kit (ASK) Generator Sets DOD Model NEP-701A each mounted on a single modified M116A2 trailer. These generator sets are air-cooled, diesel engine-driven units each with a load capacity of 3 kW at 69 Hz. Both PP trailers are two-wheeled units modified to carry an approximate load of 3200 pounds. The AN/MJQ-32 PP has its two ASK generator sets mounted on the fender/fender extensions of a modified M116A2 chassis (considered to be a nonstandard trailer). The AN/MJQ-33 PP has its dual ASK generator sets mounted on a flatbed cargo body of a modified M116A2 chassis (considered to be a Standard trailer). The main difference between the two PP's is the storage rack assembly of the AN/KQ-32 which is used to carry the antenna mast, mast kit, and mast extension kit (the AN/MJQ-33 does not contain a storage rack) and tarpaulin and bows used on the AN/MJQ-33 (the AN/MJQ-32 does not contain a tarpaulin and bows). Output from either PP is supplied to the system or equipment being powered through a 5-wire configuration switchbox located on each PP.

1-8. TABULATED DATA.

The tabulated data provides operator and unit level personnel with the dimensions and weights for power plants AN/KJQ-32 and AN/&JQ-33. These specifications are computed from the combined dimensions and weights of the generator sets and trailers as modified for use with the power plants. Specifications of the individual components can be found in their respective technical publications. For additional information concerning the ASK Generator Set, DOD Model MEP-701A, refer to TM5-6115-615-12 and -24P. For additional information on the MU16A2 trailer refer to TM9-2330-202-14&P. The tabulated data also includes the location and content of all data plates unique to the power plants.

a. Identification and Information Plates.

(1) AN/MJQ-32 power plant identification plate is located on rear section of curbside fender.

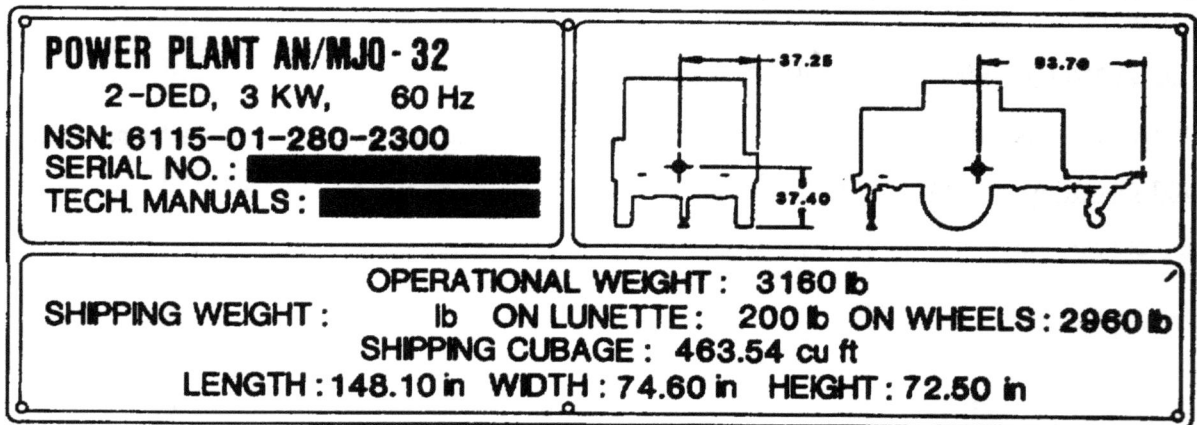

Figure 1-1. Identification Plate on AN/MJQ-32.

1-2

(2) AN/KJQ-33 power plant identification plate is located on rear section of curbside fender.

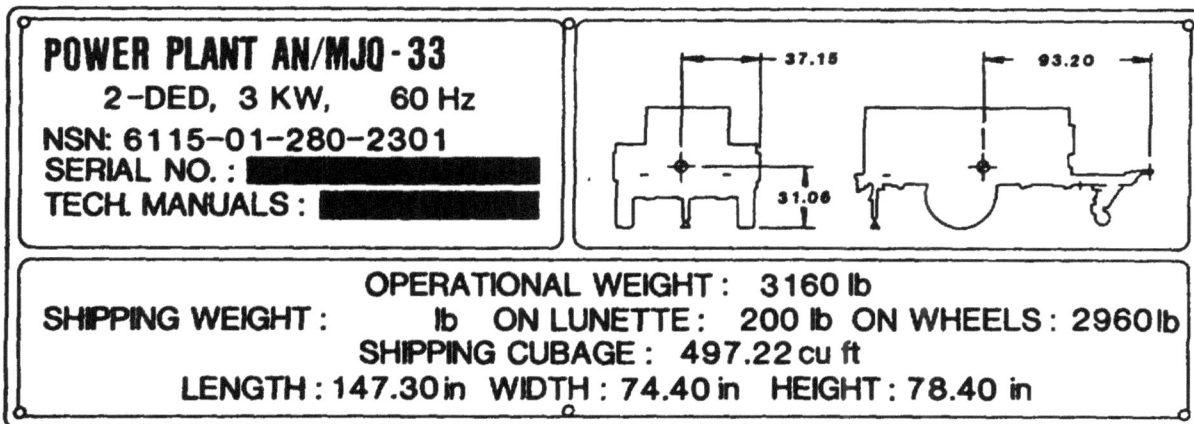

POWER PLANT AN/MJQ-33
2-DED, 3 KW, 60 Hz
NSN: 6115-01-280-2301
SERIAL NO. :
TECH. MANUALS :

OPERATIONAL WEIGHT : 3160 lb
SHIPPING WEIGHT : lb ON LUNETTE : 200 lb ON WHEELS : 2960 lb
SHIPPING CUBAGE : 497.22 cu ft
LENGTH : 147.30 in WIDTH : 74.40 in HEIGHT : 78.40 in

Figure 1-2. Identification Plate on AN/MJQ-33.

(3) Ground Terminal Identification plate.

　　(a) Location. Plates, two each for the AN/KJQ-32 and two each for the AN/MJQ-33, are located as follows:

　　　　AN/MJQ-32 AN/MJQ-33

　　　　1. Front, curbside corner　　1. Front, curbside corner

　　　　2. Rear, roadside corner　　　　2. Roadside, top of fender
　　　　　 of fender extension

　　(b) Content. GROUND TERMINAL

(4) Wiring diagram information plate on switch box (same for both power plants).

　　(a) Location. This plate is mounted inside the switch box.

　　(b) Content (see figure 4-4).

(5) Warning plate on switch box (both power plants).

(a) Location. Top of switch box.

(b) Content.

DANGER

HIGH VOLTAGE

(6) AC Ground Information plate on switch box (both power plants).

(a) Location. Plate is located on right side of switch box next to ground terminal E2.

(b) Content. AC GROUND

(7) Equipment Ground information plate on switch box (both power plants).

(a) Location. This plate is located on front of switch box.

(b) Content.

EQUIPMENT

(FRAME)

GROUND

b. Tabulated Data for Power Plants.

AN/MJQ-32 AN/MJQ-33

Overall length 148.10 in. (379.7 cm) 147.30 in. (377.7 cm)

Overall width 74.60 in. (191.3 cm) 74.40 in. (190.8 cm)

Overall height 72.50 in. (185.9 cm) 78.40 in. (201.0 cm)
                      (Top of Rack) (Top of Tarpaulin)

Net weight (empty) 3160 lb (1436.4 kg) 3160 lb (1436.4 kg)

Net weight (filled) 3228 lb (1467.3 kg) 3228 lb (1467.3 kg)

1-9. DIFFERENCES BETWEEN MODELS.

The AN/MJQ-32 and AN/MJQ-33 are identical in performance characteristics, both models consisting of two each MEP-701A, 3 kW, 60 Hz Acoustic Suppression Kit (ASK) generators. The AN/MJQ-32, however, has been fitted with an equipment rack which allows it to carry antenna masts and accessories. Also, the AN/MJQ-32 does not contain a tarpaulin and support as does the AN/MJQ-33.

Section III. PRINCIPLES OF OPERATION

1-10. PRINCIPLES OF OPERATION.

Each Power Plant, AN/MJQ-32 or AN/MJQ-33, provides 3 kW, 60 Hz precise power from either one of two modified generator sets. Power to the load equipment can be obtained through switch box terminals or directly from the generators. The trailer can be towed over prepared roads at a maximum speed of 55 mph (88.5 km/h), and over unimproved roads at a maximum speed of 30 mph (483. km/h).

1-11. LOCATION AND DESCRIPTION OF MAJOR COMPONENTS.

Refer to figures 1-3 and 1-4 for the location of major components of the AN/MJQ- 32 and AN/MJQ-33. For components not discussed in this paragraph, refer to TM5- 6115-615-12 for the generator set components and TM9-2330-202-14&P for components of the trailer.

  a. AN/MJQ-32 Power Plant (See Figure 1-3.)

      (1) STOWAGE RACK. Provides storage of antenna masts, supports, and accessories whenever power plant is moved from site to site.

      (2) GENERATOR SET (REAR). One of two Acoustic Suppression Kit (ASK) generators which provide 3 kW, 60 Hz to load equipment.

      (3) SWITCH BOX Channels power from one of two generator sets to load equipment, depending on switch setting.

      (4) FUEL CAN BRACKET. Provides mounting for fuel can during transportation. Two each.

      (5) FENDER/FENDER EXTENSION. Provides support and mounting for major components of power plant.

      (6) FIRE EXTINGUISHER BRACKET. Provides mounting for 5-pound fire extinguisher during transportation of power plant.

      (7) ANTENNA MAST MOUNT. Provides support for antenna mast during transportation of power plant. Two each.

      (8) DRIVER/PULLER HOLDER. Provides mounting for driver/puller.

      (9) ANTENNA MAST SUPPORT. Provides additional support for antenna mast during transportation of Power Plant. Two each.

      (10) FUEL CAN BRACKET. See item 4.

      (11) LOAD CABLE BRACKET. Supports load-cable reel.

      (12) FUEL-ADAPTER HOLDER. Provides mounting for drum adapter.

Change 2 1-5

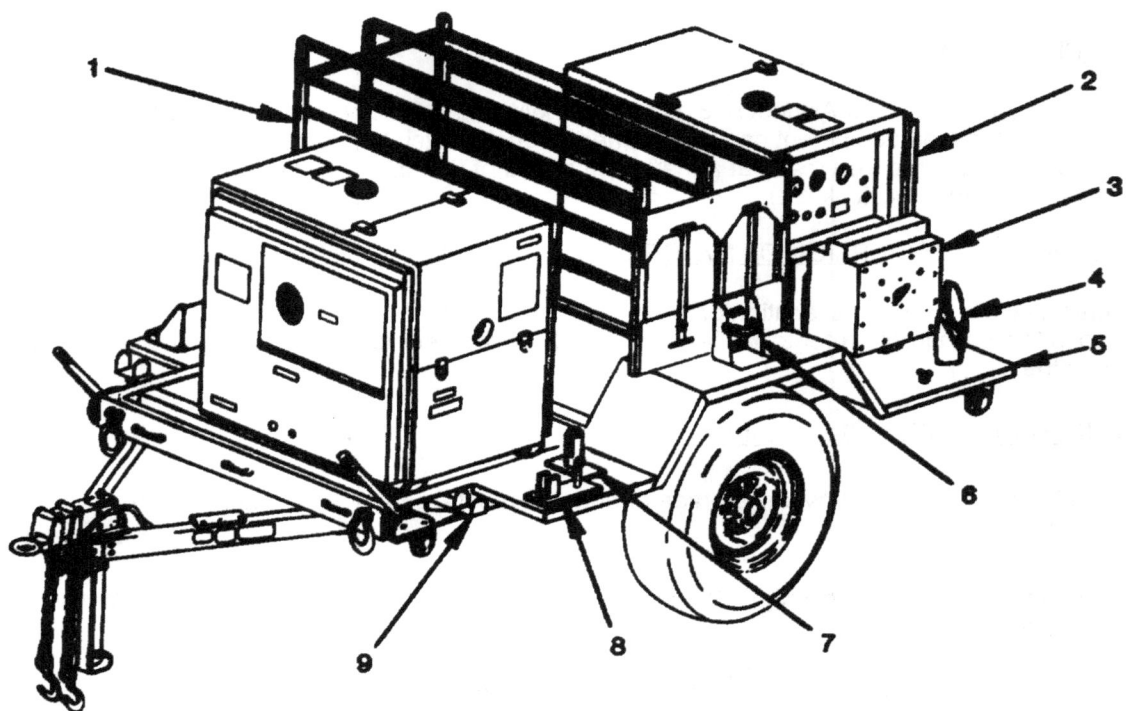

ROADSIDE FRONT

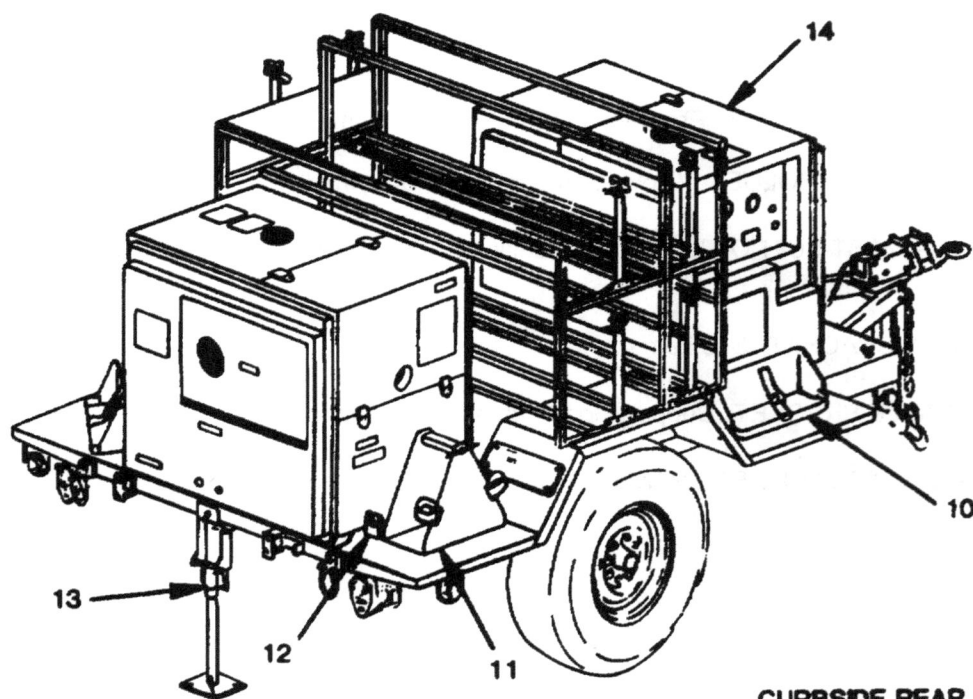

CURBSIDE REAR

Figure 1-3. AN/MJQ-32 Three-Quarter Views.

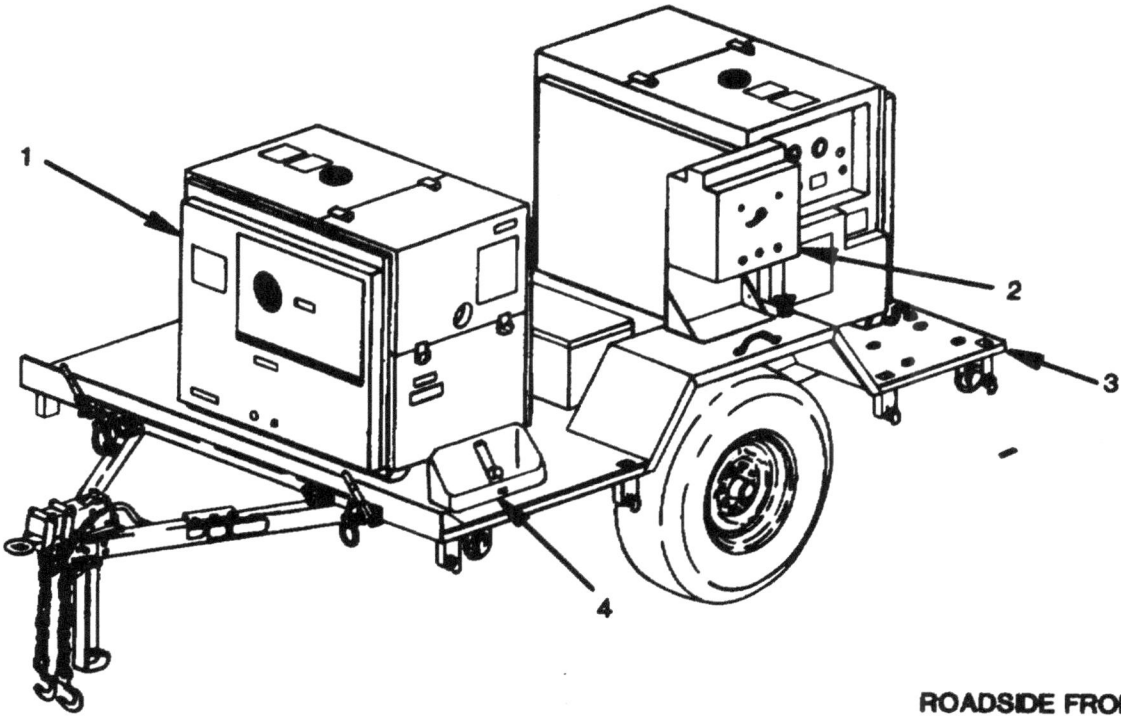

ROADSIDE FRONT

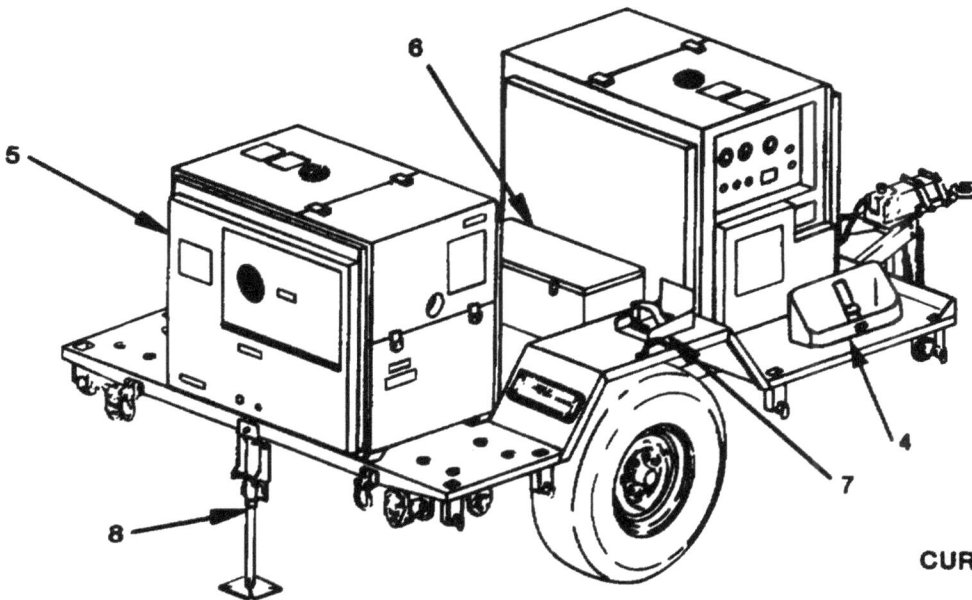

CURBSIDE REAR

Figure 1-4. AN/MJQ-33 Three-Quarter Views.

(13) REAR LEG PROP ASSEMBLY. Supports and helps stabilize power plant during operation.

(14) GENERATOR SET (FRONT). Same as item 2.

b. AN/MJQ-33 Power Plant (See Figure 1-4.)

(1) GENERATOR SET (FRONT). One of two Acoustic Suppression Kit (ASK) generators which provide 3 kW, 60 Hz to load equipment.

(2) SWITCH BOX. Channels power from one of two generator sets to load equipment, depending on switch setting.

(3) TRAILER BED. Provides support and mounting for major components of power plant.

(4) FUEL CAN BRACKET. Provides mounting for fuel can during transportation of power plant. Four each.

(5) GENERATOR SET (REAR). Same as item (1).

(6) ACCESSORY BOX. Provides stowage for certain basic items (hammer, ground rods, etc.) during transportation or operation of power plant.

(7) FIRE EXTINGUISHER BRACKET. Provides mounting for 5-pound fire extinguisher during transportation of power plant.

(8) REAR LEG PROP ASSEMBLY. Supports and helps stabilize power plant during operation.

Change 2   1-8

CHAPTER 2
OPERATING INSTRUCTIONS

Section I. OPERATING PROCEDURES

2-1. <u>OPERATING PROCEDURES.</u>

Before the generators of either power plant are started and operated, the power plant is towed to a worksite and installed. Operating procedures are the same for both the AN/MJQ-32 and AN/MJQ-33.

   a. <u>Generator Set Operating Procedures.</u> <u>Detailed prestarting, start up, operating, and shutdown procedures for the</u> generator sets can be found on the Operating Instructions data plate located on the left-hand side of the front panel of each ASK generator set and in the generator set technical manual TM5-6115-615-12.

WARNING
Do not operate power plant until it is properly grounded. Serious injury or death by electrocution can result from operating an ungrounded generator set.

CAUTION
To avoid damage to equipment, make certain of voltage, frequency, and phase requirements of load connected to generator sets.

CAUTION
Make sure generator set circuit breakers and switch box rotary switch are in OFF position before proceeding. Damage to equipment may result.

b. <u>Switch Box Operating Procedures.</u>

   (1) Single generator set operation. Use switch box to operate only one generator set as follows: (See figure 2-1).

      (a) Make sure power and ground cables are connected to generator set and switch box.

      (b) Start generator and bring it up to rated speed, voltage, and frequency. (Refer to TM5-6115-615-12).

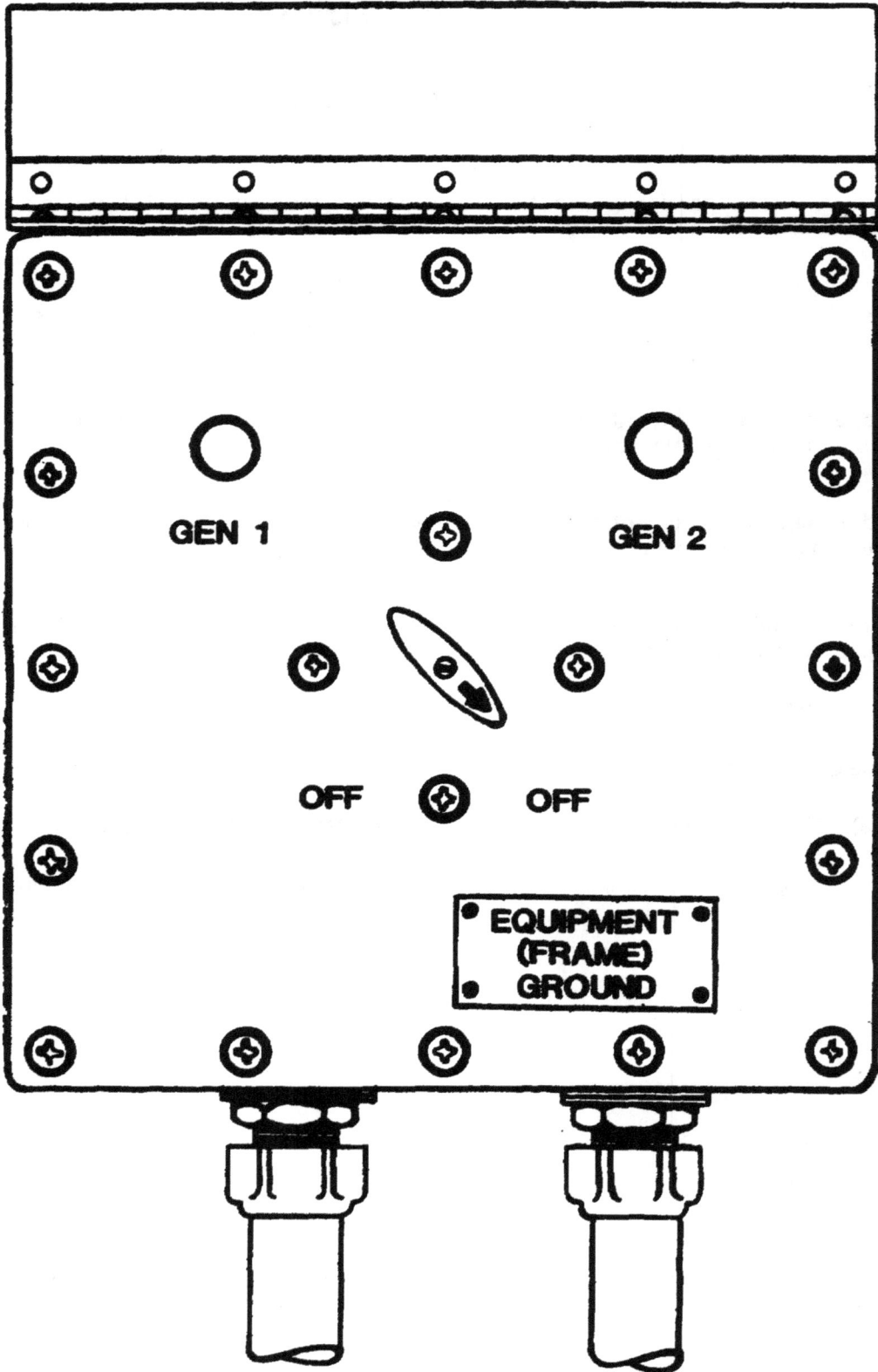

Figure 2-1. Switch Box Control Panel.

(c) Set generator circuit breaker to ON position.

(d) Turn rotary switch on switch box to GEN 1 or GEN 2 position, as applicable.

(e) To stop operation, move rotary switch and generator set circuit breaker to OFF position. Shut down generator set. (Refer to TM5-6115-615-12).

(2) Dual generator set operation. Use the switch box to alternately operate both generator sets as follows:

(a) Make sure power and ground cables are connected to switch box and both generator sets.

(b) Select first generator set to be used and bring it into operation in accordance with paragraph 2-1b(1), steps (b) thru (d).

(c) Start second generator set and bring it up to rated speed, voltage and frequency. (Refer to TM5-6116-615-12).

NOTE
Both generator set lights on switch box panel will be on whenever both generator sets are running and providing power to load, regardless of rotary switch setting.

(d) Move switch box rotary switch to GEN position corresponding to second generator set.

(e) Set circuit breaker on first generator set to OFF position and shut down generator set. (Refer to TM5-6115-615-12).

(f) To stop operation, move rotary switch to OFF position and open circuit breaker on generator set still running. Shut down generator set. (Refer to TM5-6115-615-12).

c. Trailer Operating Procedures. Refer to TM9-2330-202-14&P for specific operating procedures for the M116A2 trailer.

Section II. OPERATION UNDER UNUSUAL CONDITIONS

2-2. OPERATION UNDER UNUSUAL CONDITIONS.

When operating the power plant under unusual conditions such as extremes in temperature or difficult terrain, there are steps that must be taken to protect the equipment.

   a. Refer to TM5-6115-615-12 for specific procedures when operating the ASK generator sets under unusual conditions.

   b. Refer to TM9-2330-202-14&P for specific procedures when operating the trailers under unusual conditions.

Section III. OPERATION OF AUXILIARY EQUIPMENT

2-3. OPERATION OF AUXILIARY EQUIPMENT.

There is no auxiliary equipment supplied with the power plants.

Section IV. PREVENTIVE MAINTENANCE CHECKS AND SERVICES

2-4. GENERAL.

The operator/crew preventive maintenance checks and services (PMCS) listed in table 2-1 are grouped according to stages of equipment operation or time intervals. Using the following as a guide, do the checks and services at the intervals shown. All PMCS pertain to both the AN/MJQ-32 and AN/NJQ-33 unless otherwise noted.

   • Before you operate, perform your before (B) PMCS. Observe all CAUTIONS and WARNINGS.

   • While you operate, perform your during (D) PMCS. Observe all CAUTIONS and WARNINGS.

   • After you operate, be sure to perform your after (A) PMCS.

   • Do (W) PMCS weekly.

   • Do (M) PMCS monthly.

   • If equipment fails to operate, refer to Section IV Troubleshooting. If the problem cannot be corrected, refer to DA Pam 738-750.

a. Purpose of PMCS Table. The purpose of the PMCS table is to provide a systematic method of inspecting and servicing the equipment. In this way, small defects can be detected early before they become a major problem causing the equipment to fail. The PMCS table is arranged with the individual PMCS procedures listed in sequence under assigned intervals. The most logical time (before, during, or after operation) to perform each procedure determines the interval to which it is assigned. Make a habit of doing the checks and services in the same order each time and anything wrong will be seen quickly. See paragraph below for an explanation of the columns in table 2-1.

b. Explanation of Columns. The following is a list of the PMCS table column headings with a description of the information found in each column.

(1) ITEM NO column. Provides identification numbers for the functions to be made. It should be used as a reference on reports of failures or deficiencies.

(2) INTERVAL column. Identifies when a particular function (check/service) is to be performed, that is; before, during, or after operation.

(3) ITEM TO BE INSPECTED - PROCEDURE column. Identifies the item to be inspected (checked/serviced). The procedure(s) to be followed is shown (indented) below.

(4) EQUIPMENT IS NOT READY/AVAILABLE IF column. Tells you when and why your equipment cannot be used.

NOTE
The terms ready/available and mission capable refer to the same status: Equipment is on hand and is able to perform its combat mission.

c. Reporting. - "If your equipment does not perform as required, refer to Chapter 3, Maintenance Instructions." Report any malfunctions or failures on DA Form per DA Pamphlet 738-750.

2-5. SPECIAL INSTRUCTIONS.

Preventive maintenance is not limited to performing the checks and services listed in the PMCS table. Covering unused receptacles, stowing unused equipment and other routine procedures such as equipment inventory, cleaning components, and touch-up painting are not listed in the PMCS table. These are things you should do any time you see they need to be done. If a routine check is listed in the PMCS table, it is because other operators have reported problems with this item. Take along tools and cleaning cloths needed to perform the required checks and services. Use the information in following paragraphs to help you identify problems at any time.

a. Routine Inspections. Use the following information to help identify potential problems before and during checks and services.

WARNING

Clean parts in a well-ventilated area. Avoid inhalation of solvent fumes and prolonged exposure of skin to cleaning solvent. Wash exposed skin thoroughly. Dry cleaning solvent (PD-680) used to clean parts is potentially dangerous to personnel and property. Do not smoke or use near open flame or excessive heat. Flash point of solvent is 100°F to 138°F (38°C to 590C).

(1) Keep it clean. Dirt, grease, and oil get in the way and may cover up a serious problem. Use drycleaning solvent PD-680 (item 1, appendix F), to clean metal surfaces. Use soap and water to clean rubber or plastic parts and material.

(2) Bolts, nuts, and screws. Check them all to make sure they're not loose, missing, bent, or broken. Don't try to check them all with a tool, but look for chipped paint, bare metal, or rust around bolt heads. If you find one loose, tighten it or report it to unit level maintenance.

(3) Welds. Look for loose or chipped paint, rust, or gaps where parts are welded together. If a broken weld is found, report it to unit level of maintenance.

(4) Electrical wires, connectors, terminals, and receptacles. Look for cracked or broken insulation, bare wires, and loose or broken connectors. Tighten loose connectors and make sure the wires are in good condition. Examine terminals and receptacles for serviceability. If deficiencies are found, report them to unit level maintenance.

(5) Hoses and fluid lines. Look for wear, damage, and leaks. Make sure clamps and fittings are tight. Wet spots and stains around a fitting or connector can mean a leak. If a leak comes from a loose connector, tighten it. If something is broken or worn out, report it to unit level maintenance.

b. Leakage Definitions. It is necessary for you to know how fluid leakage affects the status of your equipment. The following are definitions of the types/classes of leakage you need to know to be able to determine the status of your equipment. Learn and be familiar with them. When in doubt, NOTIFY YOUR SUPERVISOR.

Leakage Definitions:

Class I Seepage of fluid (as indicated by wetness or discoloration) not great enough to form drops.

Class II Leakage of fluid great enough to form drops but not enough to cause drops to drip from item being checked/inspected.

Class III Leakage of fluid great enough to form drops that fall from the item being checked/inspected.

CAUTION

Equipment operation is allowable with minor leakage (Class I or II) of any fluid except fuel. Of course, consideration must be given to the fluid capacity in the item being checked/inspected.

When in doubt, notify your supervisor. When operating with Class I or II leaks, continue to check fluid level more often than required in the PMCS. Parts without fluid will stop working and/or cause equipment damage. Class III leaks should be reported to your supervisor or unit level maintenance.

NOTE

If the equipment must be kept in continuous operation, check and service only those items that can be checked and serviced without disturbing operation. Make the complete checks and services when the equipment can be shut down. Within designated interval, these checks are to be performed in the order listed.

Table 2-1. Operator/Crew Preventive Maintenance Checks and Services

B - Before D - During A - After W - Weekly M - Monthly

ITEM NO.	INTERVAL					Item to be inspected Procedure: check for and have repaired, filled, or adjusted as needed	Equipment is not ready/available if:
	B	D	A	W	M		
						WARNING Before performing anymaintenance that requiresclimbing on or under-trailer, set trailerhandbrakes, chock wheels,and lower rear leg prop. Injury to personnel couldresult from trailer sud-denly rolling or tipping. **NOTE** Perform weekly as well asbefore PMCS if: You are the assignedoperator but have notoperated the equipmentsince the last weeklyinspection. You are operating theequipment for the first time. Generator set checks andservices in this table aredescribed as performed ona single generator set. These procedures must beperformed on each of thegenerator sets that make upthe AN/MJQ-32 and AN/MJQ-33. Generators are shown with ASK panel(s) removed. For panel removal referto TM5-6115-615-12.	

Table 2-1. Operator/Crew Preventive Maintenance Checks and Services - Continued

B - Before D - During A - After W - Weekly M - Monthly

ITEM NO.	INTERVAL					Item to be inspected Procedure: check for and have repaired, filled, or adjusted as needed	Equipment is not ready/available if:
	B	D	A	W	M		
1						GENERATOR SET EXTERIORa. Check on, around, and beneath the generator set (1) for fuel or oil leaks. b. Check that generator set grounds (2) are properlyinstalled, and groundingconnections are tight. 	Class III lubrication oil or any class fuel leak is detected. Not properly grounded.
2						ENGINE OIL LEVEL Open service access door, Panel' 5, and check oil filter dip-stick (3) for proper oil level Add oil as required. 	

Table 2-1. Operator/Crew Preventive Maintenance Checks and Services - Continued

B - Before D - During A - After W - Weekly M - Monthly

| ITEM NO. | INTERVAL | | | | | Item to be inspected Procedure: check for and have repaired, filled, or adjusted as needed | Equipment is not ready/avail-able if: |
	B	D	A	W	M		
3						AIR CLEANER INDICATOR Remove Panel 4 and check indi-cator (4) for a restricted aircleaner. If red warn-ing indi-cator becomes visible, notifyunit maintenance for cleaningor replacement.	
4						ACCESSORIES Check that the following acces-sories are not missing ordamaged from each power plant.a. Sledge hammersb. Fire extin-guisher's c. Gas cansd. Can spoutse. Driver/pullerf. Ground rods	Fire extinguisher ismissing, damaged, orseal is broken. Ground rods are missingor unserviceable.

Table 2-1. Operator/Crew Preventive Maintenance Checks and Services - Continued

B - Before D - During A - After W - Weekly M - Monthly

ITEM NO.	INTERVAL					Item to be inspected Procedure: check for and have repaired, filled, or adjusted as needed	Equipment is not ready/available if:
	B	D	A	W	M		
5						ACCESSORIES - CONTg. Gas can adaptersh. Cable reel (AN/MJQ-32) BRACKETS/HOLDERS/MOUNTS/ GROUND STUDS Check fire extinguisher, fuelcan, switch box mounting brack-ets and ground studs for loosehardware and broken fittings on-both power plants. Check thecable reel and hammer brackets,spout and driver/ puller mountand mast support, all on the AN/MQ-32 .	Ground studs are missing or unserviceable.
6		•		•		TIRESa. Check tires (5) for cuts, foreign objects, or unusualtread wear. Remove anystones from between thetreads.	One tire if flat, miss- ing, or unserviceable.

Table 2-1. Operator/Crew Preventive Maintenance Checks and Services - Continued

B - Before D - During A - After W - Weekly M - Monthly

ITEM NO.	INTERVAL					Item to be inspected Procedure: check for and have repaired, filled, or adjusted as needed	Equipment is not ready/available if:
	B	D	A	W	M		
7						TIRES - CONTb. Check that tire pressure is 35 PSI (241.22 kPa) whentires are cool. WHEELS Check for damage and for miss-ing or loose stud nuts (6).	One or more wheels are damaged. Two or morestud nuts are loose or missing.
8						DRAWBAR RING Check drawbar ring (7) for in- secure mounting and obvious damage.	Ring is loose or bent.
9						INTERVEHICULAR CABLE Check cable (8) and connectorfor cuts and breaks.	Intervehicular cable is broken or missing.
10		•		•	•	SAFETY CHAINS Check safety chains (9) for in-secure mounting and obviousdamage.	Safety chains are miss- ing, or unsecured.

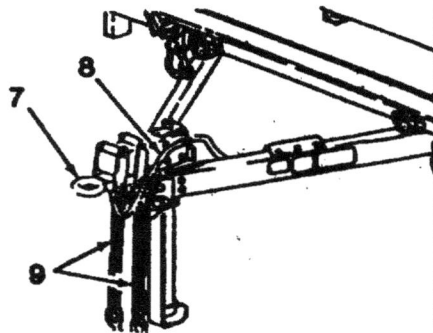

Table 2-1. Operator/Crew Preventive Maintenance Checks and Services - Continued

B - Before D - During A - After W - Weekly M - Monthly

ITEM NO.	INTERVAL					Item to be inspected Procedure: check for and have repaired, filled, or adjusted as needed	Equipment is not ready/available if:
	B	D	A	W	M		
11						BOW ASSEMBLIES AND TARPAULIN SUPPORT (AN/MJQ-33 ONLY) Inspect four bow assemblies (10)and tarpaulin support (11).	
12						TARPAULIN COVER (AN/MJQ-33 ONLY)	
						a. Check tarpaulin cover (12) for missing and defectivetiedown straps and snapfasteners (13).	
						b. Check for missing and defective ropes (14).	
						c. Check for missing and defective straps and buckles (15).	
						d. Check for ripped seams and tears.	

Table 2-1. Operator/Crew Preventive Maintenance Checks and Services - Continued

B - Before D - During A - After W - Weekly M - Monthly

ITEM NO.	INTERVAL					Item to be inspected Procedure: check for and have repaired, filled, or adjusted as needed	Equipment is not ready/available if:
	B	D	A	W	M		
13						LIGHTSa. With intervehicular cable connected to towing vehi-cle, operate vehicle lightswitch through all setting-sand check lights.	Lights inoperative or unserviceable.
						NOTE An assistant is requiredwhile checking brake lights.	
						b. Step on brake pedal of tow- ing vehicle and check brake lights (16)	Brake lights inopera- tive or unserviceable.

Table 2-1. Operator/Crew Preventive Maintenance Checks and Services - Continued

B - Before   D - During   A - After   W - Weekly   M - Monthly

ITEM NO.	INTERVAL					Item to be inspected Procedure: check for and have repaired, filled, or adjusted as needed	Equipment is not ready/available if:
	B	D	A	W	M		
14						SUPPORT LEG ASSEMBLY With trailer connected to towing vehicle, check support leg assembly (17) for ease of operation.	upport leg assembly is seized. Leg prop assembly is unserviceable.
15						REAR LEG PROP ASSEMBLY Inspect leg prop assembly (18) for broken or missing parts	

Table 2-1. Operator/Crew Preventive Maintenance Checks and Services - Continued

B - Before D - During A - After W - Weekly M - Monthly

ITEM NO.	INTERVAL					Item to be inspected Procedure: check for and have repaired, filled, or adjusted as needed	Equipment is not ready/avail-able if:
	B	D	A	W	M		
16						BRAKE SYSTEM Test brake system by hookingtrailer to towing vehicle andapplying brakes.	Service brakes fail to operate.

17						TRAILER OPERATIONa. Be alert for any
						
						unusual
						    noises while towing trailer.
						    Stop and investigate anyunusual
						    noises.
						
						b. Ensure that trailer is
						    tracking/following correct1:behind
						    towing vehicle withno side pull.	

18			•			GENERATOR SET GAGES AND
						
						INSTRUMENTSa. Check that frequency
						(19)
						    indicates 60 Hz (red line)when gener-
						    ator is operatingunder load.
							Correct frequency can- not be maintained.

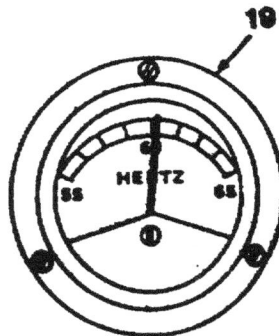

Table 2-1. Operator/Crew Preventive Maintenance Checks and Services - Continued

B - Before D - During A - After W - Weekly M - Monthly

ITEM NO.	INTERVAL					Item to be inspected Procedure: check for and have repaired, filled, or adjusted as needed	Equipment is not ready/avail-able if:
	B	D	A	W	M		
18						GENERATOR SET GAGES AND INSTRUMENTS - CONTb. Check that current meter (20) reading does not ex-ceed 100 percent or morethan 5 percent load differ-ence between phases.	eter reading exceeds 100 percent or more than 5 percent load difference between phases.

20

PERCENT LOAD CURRENT

. Check that voltmeter (21) indicates desired outputvoltage as determined byload connections and amps-volts transfer switch.

Desired voltage cannot be obtained and main- tained.

21

A. C. VOLTS

Table 2-1. Operator/Crew Preventive Maintenance Checks and Services - Continued

B - Before D - During A - After W - Weekly M - Monthly

ITEM NO.	INTERVAL					Item to be inspected Procedure: check for and have repaired, filled, or adjusted as needed	Equipment is not ready/available if:
	B	D	A	W	M		
19						FUEL TANK	

.. Remove Panel 4 and fill
   tank (22) upon completionof operation.

NOTE

Fuel system temperature must beabove freezing when drainingwater and sediment.b. Open drain (23) and drain water and sediment fromfuel tank. Allow to drainuntil fuel runs clean.

Table 2-1. Operator/Crew Preventive Maintenance Checks and Services - Continued
B - Before D - During A - After W - Weekly M - Monthly

ITEM NO.	INTERVAL					Item to be inspected	Equipment is not ready/available if:
	B	D	A	W	M	Procedure: check for and have repaired, filled, or adjusted as needed	
20			●			FUEL STRAINER AND FILTERS Drain water and sediment fromfuel filter assembly (24). Allow to drain until fuel runs clean.	
21			●			HANDBRAKES With trailer hooked to towingvehicle, set handbrakes (25). Move trailer slightly to see ifhand-brakes hold wheels.	

Table 2-1. Operator/Crew Preventive Maintenance Checks and Services - Continued
B - Before D - During A - After W - Weekly M - Monthly

ITEM NO.	INTERVAL					Item to be inspected / Procedure: check for and have repaired, filled, or adjusted as needed	Equipment is not ready/available if:
	B	D	A	W	M		
21			●			HANDBRAKES - CONT	
22			●			BRAKE DRUMS AND HUBS **WARNING** **A defect in the operation of thebrakes or hub can cause these partsto get hot enough to cause seriousburns. Use extreme caution whenattempting to detect heat in this area.** Feel for overheating to detectdragging or binding	Brakes or hub are dragging or binding.
23				●		REFLECTORS Check for damaged or missingreflectors.	

Table 2-1. Operator/Crew Preventive Maintenance Checks and Services - Continued
B - Before D - During A - After W - Weekly M - Monthly

ITEM NO.	INTERVAL					Item to be inspected Procedure: check for and have re-paired, filled, or adjusted as needed	Equipment is not ready/available if:
	B	D	A	W	M		
24				●		BATTERIES Check battery (26) electrolytelevel. Level should be about 3/4 inch above top of plates. If water level is low notifyunit level maintenance.	
25				●		TRAILER FRAME Inspect entire chassis framefor dam-age, cracks, and broken welds.	Frame is broken or cracked.
26	●					SWITCH BOX Inspect for broken, damaged,or loose connectors; brokeninsulation, frayed or brokenwires. Inspect rotary switch-for correct operation.or has missing parts.	Connectors are broken, damaged or loose, bare wires are exposed; rotary switch is defec- tive, not operational,

CHAPTER 3
OPERATOR/CREW MAINTENANCE INSTRUCTIONS

Section I. LUBRICATION INSTRUCTIONS

3-1. GENERAL.
Detailed instructions for the lubrication of the major components of the power plants are contained in the applicable Lubrication Orders (LO's).

3-2. GENERATOR LUBRICATION.
Refer to TM5-6115-615-12 for generator set lubrication instructions.

3-3. TRAILER LUBRICATION
There are no operator/crew lubrication requirements for the power plant trailers. However, the operator shall assist unit maintenance.

Section II. COMPONENTS OF END ITEM LIST (COEIL), ADDITIONAL AUTHORIZATION

LIST (AAL), AND EXPENDABLE/DURABLE SUPPLIES AND MATERIALS LIST (E/DS & ML)

3-4. GENERAL.
See Appendix D for COEIL, Appendix E for AAL, and Appendix F for E/DS & ML.

Section III. TROUBLESHOOTING

3-5. POWER UNIT TROUBLESHOOTING.
There are no troubleshooting procedures authorized at operator level for the power plant end item. Troubleshooting procedures for the individual generator sets and trailer are contained in their respective technical manuals referenced below:

    a. Generator Set Troubleshooting. Refer to TM5-6115-615-12 for troubleshooting procedures.

    b. Trailer Troubleshooting. Refer to TM9-2330-202-14&P for troubleshooting procedures.

Section IV. OPERATOR/CREW MAINTENANCE

3-6. TARPAULIN SUPPORT AND BOW ASSEMBLY MAINTENANCE.
Maintenance of the tarpaulin support and bow assembly at operator level is limited to replacement of the tarpaulin support and/or the bow assembly both of which are part of the AN/MJQ-33 only.

   a. Tarpaulin Replacement. (See Figures 3-1 and 3-2.)

      (1) Removal.

         (a) Untie 25 ropes (1, figure 3-1) fastening tarpaulin to trailer body (2).
         (b) Unfasten six straps and buckles (3) securing rear curtain (4). Roll up curtain, and secure with three roll-up straps (5) provided.
         (c) Unfasten six straps and buckles (3) securing front curtain (6). Roll up curtain, and secure with three roll-up straps (5) provided.
         (d) Roll up each side (7) of tarpaulin, in turn, and secure each side with four roll-up straps (8) provided.
         (e) Working under tarpaulin (1, figure 3-2), unfasten eight straps (2) securing tarpaulin to bow assemblies (3). Remove tarpaulin.
      (2) Installation.

NOTE
Front curtain is provided with three tie-down ropes. Rear curtain has only two tie-down ropes.

         (a) Position tarpaulin (1, figure 3-2) on top of bows (3) making certain front of tarpaulin is at front of trailer.
         (b) Secure tarpaulin (1) to bow assembly (3) with eight straps (2) provided.
         (c) Unfasten roll-up straps (8, figure 3-1) securing sides (7) of tarpaulin and lower both sides.
         (d) Unfasten roll-up straps (5) securing front and rear curtains (4, 6) and lower both curtains.

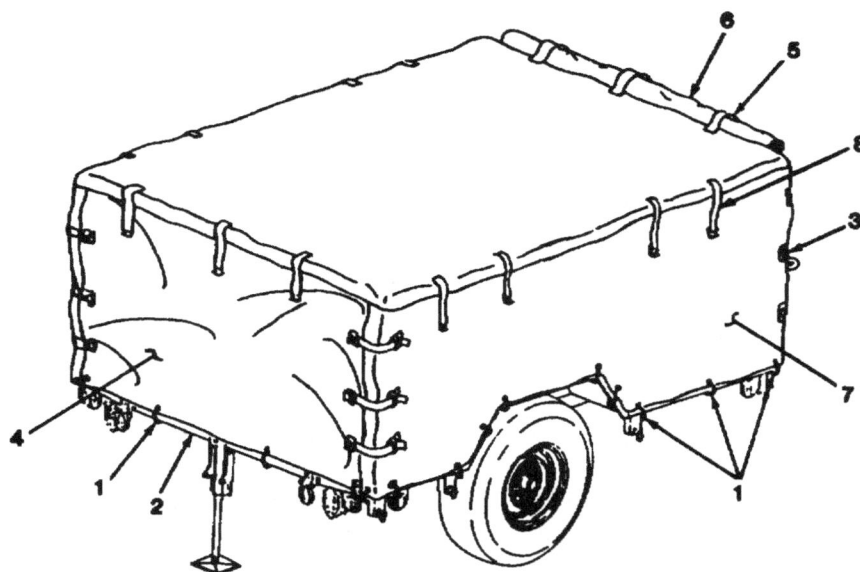

Figure 3-1. Tarpaulin Installed on AN/MJQ-33.

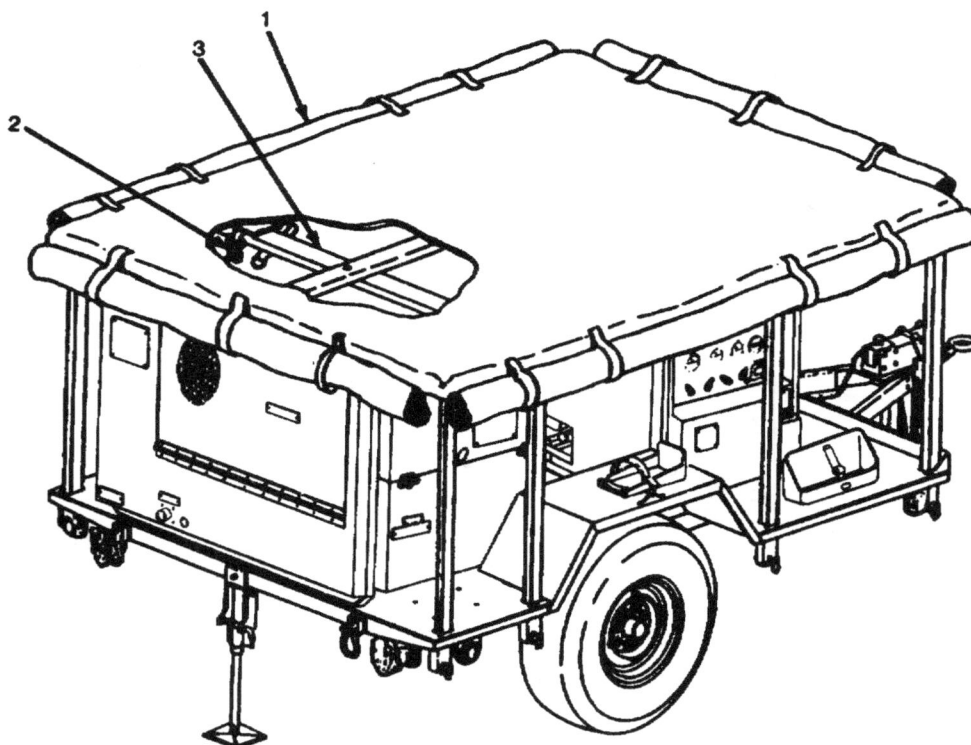

Figure 3-2. Tarpaulin Rolled Up for Removal on AN/MJQ-33.

(e) Secure front and rear curtains (4, 6) to sides (7) with six straps and buckles (3) provided on each curtain.

(f) Secure tarpaulin to trailer body (2) with 25 ropes i(1) provided.

b. Tarpaulin Support and Bow Assembly Replacement. (See Figures 3-3 and 3-4.)

(1) Removal.

(a) Remove tarpaulin (paragraph 3-6a(1)).

(b) Remove wingnut (1, figure 3-3), lockwasher (2), two flat washers (3) and screw (4) securing tarpaulin support (5) to each of four bow assemblies (6) and remove tarpaulin support.

(c) Remove two quick release pins (1, figure 3-4) securing bow assembly (2) in pocket (3) on trailer body (4). Lift each bow out of pocket and off trailer body.

(2) Installation.

(a) Lift each bow (2, figure 3-4) on trailer, align bow ends with pockets (3) in trailer body (4) and drop bow in place. Secure each bow assembly with two quick release pins (1) provided.

(b) Position tarpaulin support (5, figure 3-3) on bows (6) and secure tarpaulin support to each bow with one screw (4), two flat washers (3), lockwasher (2), and wing nut (1).

(c) Install tarpaulin on trailer (paragraph 3-6a(2)).

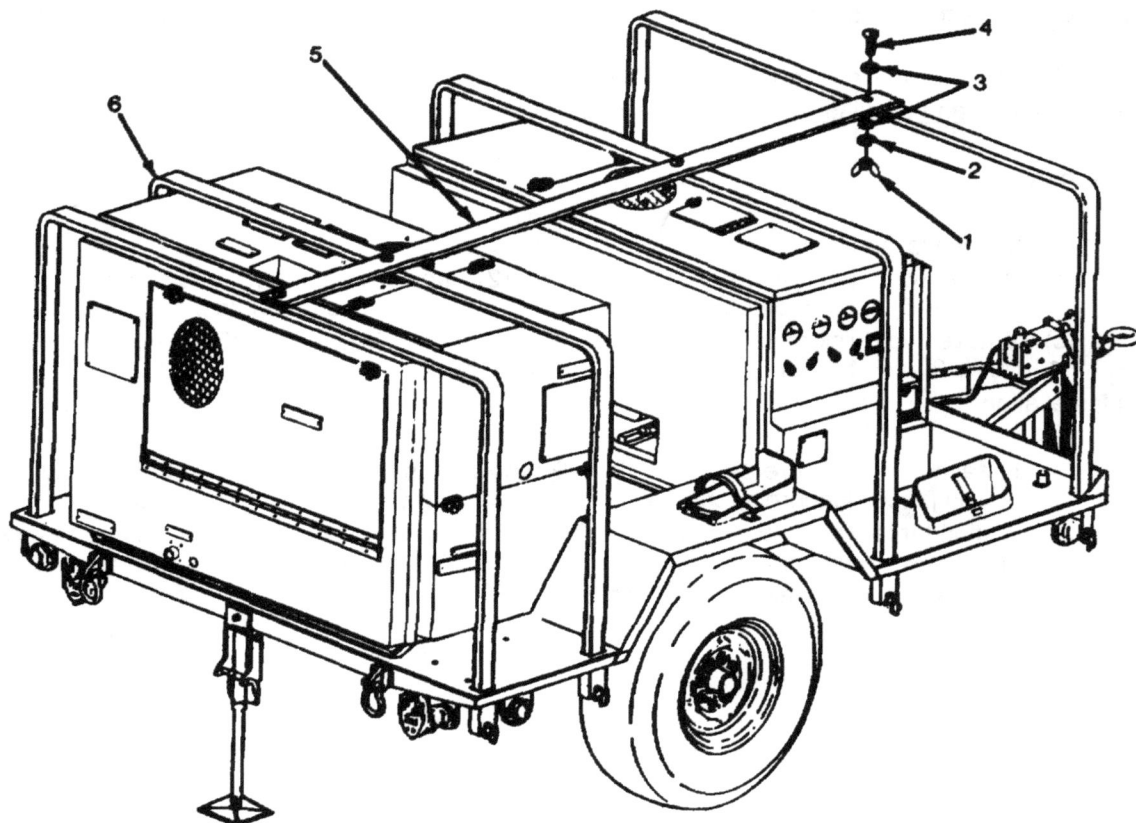

Figure 3-3. Tarpaulin Support Replacement on AN/MJQ-33.

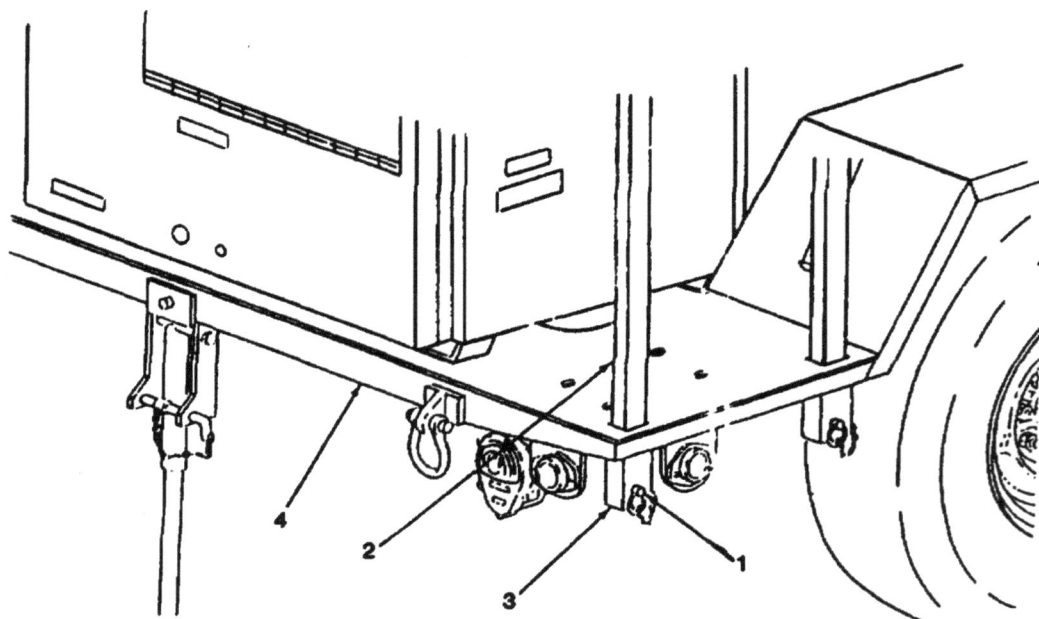

Figure 3-4. Bow Assembly Replacement on AN/MJQ-33.

3-5/(3-6 blank)

CHAPTER 4
UNIT MAINTENANCE

Section I. SERVICE UPON RECEIPT OF EQUIPMENT

4-1. INSPECTING AND SERVICING EQUIPMENT.
The power plants shall be unpacked, inspected and serviced as described in the following paragraphs. Unpacked equipment must be checked against the equipment packing list to ensure completeness. Discrepancies must be reported in accordance with instructions given in DA Pam 738-750.

a. Unpacking Power Plant AN/MJQ-32 (Overseas: Classification B/A).

WARNING
Steel strapping used in packaging of the power plant has sharp edges. Care should be taken when cutting and handling strapping to avoid injury to personnel.

WARNING
Basic Issue Items List (BIIL) box weighs approximately 125 pounds. Use at least two men when removing box from stowage rack to avoid injury to personnel.

(1) Remove and set aside packing list and shortage packing list (if applicable) from side of crate.
(2) Using metal cutters, cut the steel strapping from around the crate and remove crate from trailer.
(3) Cut the steel strappings which secure the BIIL box to the top of the stowage rack. Slide BIIL box forward and remove from rack.
(4) Cut. plastic tie-wraps securing fuel cans to interior of stowage rack and remove fuel cans.
(5) Remove canvas cover from switchbox.
(6) Uncrate the BIIL box and remove all packaging/cushioning material.
(7) Remove packaging/cushioning material from all other accessories.

(8) Using the packing list previously removed in step (1) above, inventory the items in BIIL box and all other accessories. Check missing items against shortage packing list (if sent). Report any discrepancy to your supervisor.

(9) Remove and retain all tags from components for informational purposes.

(10) Remove all plastic tie-wraps necessary to place power plant into operation.

b. Unpacking Power Plant AN/NJQ-32 (Stateside: Classification B/C).

(1) Remove and set aside the packing list and shortage packing list (if applicable) from plywood placard located in front of front generator.

(2) Unfasten stowage rack nylon strappings and remove placard.

WARNING
Steel strapping used in packaging of the power plant has sharp edges. Care should be taken when cutting and handling strapping to avoid injury to personnel.

WARNING
Basic Issue Items List (BIIL) box weighs approximately 125 pounds. Use at least two men when removing box from stowage rack to avoid injury to personnel.

(3) Perform steps (3) thru (10) in a above.

c. Unpacking Power Plant AN/MJQ-33 (Overseas: Classification B/A).

WARNING
Steel strapping used in packaging of the power plant has sharp edges. Care should be taken when cutting and handling strapping to avoid injury to personnel.

WARNING
Bow assembly will fall and cause injury to personnel if not supported before cutting away steel strappings.

(1) Remove and set aside packing list and shortage packing list (if applicable) from side of one of the generator crates.
(2) Supporting bow assembly, cut away steel strappings which secure bow assembly to trailer and to each other.
(3) Remove individual bows by sliding each one forward and out of their supporting wooden braces.
(4) Remove wooden braces from bow pockets by removing each of the quick-release pins.
(5) Using metal cutters, cut away steel strapping from rear generator crate.
(6) Remove both front and rear generator crates from trailer.
(7) Remove tarpaulin package located under one of the generator crates and remove tarpaulin from its plastic storage.
(8) Cut away steel strappings from switchbox crate and remove crate.
(9) Remove and retain all tags from components for informational purposes.
(10) Remove all plastic tie-wraps necessary to place power plant into operation.

d. Unpacking Power Plant AN/MJQ-33 (Stateside: Classification B/C).

(1) Remove packing list and shortage packing list (if applicable) from plywood placard situated in front of rolled-down tarpaulin.
(2) Cut away plastic tie-wraps and unfasten tarpaulin nylon strappings supporting placard and remove placard.
(3) Roll up tarpaulin (paragraph 3-6a(1)(a) thru (d)). Remove tarpaulin if desired (paragraph 3-6a(1)(e)).
(4) Remove and retain all tags from components for informational purposes.
(5) Remove all plastic tie-wraps necessary to place power plant into operation.

e. Inspecting Power Plants AN/I)Q-32 and AN/UMQ-33.

    (1) Refer to Service Upon Receipt of Materiel in TM5-6115-615-12 for initial inspection procedures for generator sets.
    (2) Refer to Service Upon Receipt of Materiel in TM9-2330-202-14&P for initial inspection procedures for trailer.

f. Servicing Power Plants AN/MJQ-32 and AN/MJQ-33.

    (1) Remove the depreservation guide. The depreservation guide explains what was done to the equipment prior to packaging and what has to be done to place the power plants into operation.
    (2) Refer to paragraph 4-le(1) and (2) for initial servicing procedures for the generator sets and trailer.

4-2. INSTALLATION (SEE FIGURE 4-1).

Installation of the power plants AN/MJQ-32 and AN/MJQ-33 at a worksite involves positioning the trailer and grounding the power plants. Since both power plant installation procedures are alike, only the AN/MJQ-33 will be shown in figure 4-3.

a. Positioning Power Plant. Position the power plant on the worksite as follows:

    (1) Select an area as level as possible to install power plant and position trailer.
    (2) Set trailer handbrakes and lower trailer support leg.
    (3) Chock both wheels and lower rear leg prop assembly. Adjust leg prop assembly by turning inner leg until leg base makes firm contact with ground.
    (4) Lift and secure the tarpaulin (AN/MJQ-33 only) in raised position away from generator set exhaust.

WARNING
Remove fire extinguishers and fuel cans prior to start-up of generator. This will ensure that in the event of fire extra fuel will not be involved and extinguisher will remain accessible.

    (5) Locate fuel cans and fire extinguisher on ground away from power pl ant.

(THIS PAGE INTENTIONALLY LEFT BLANK)

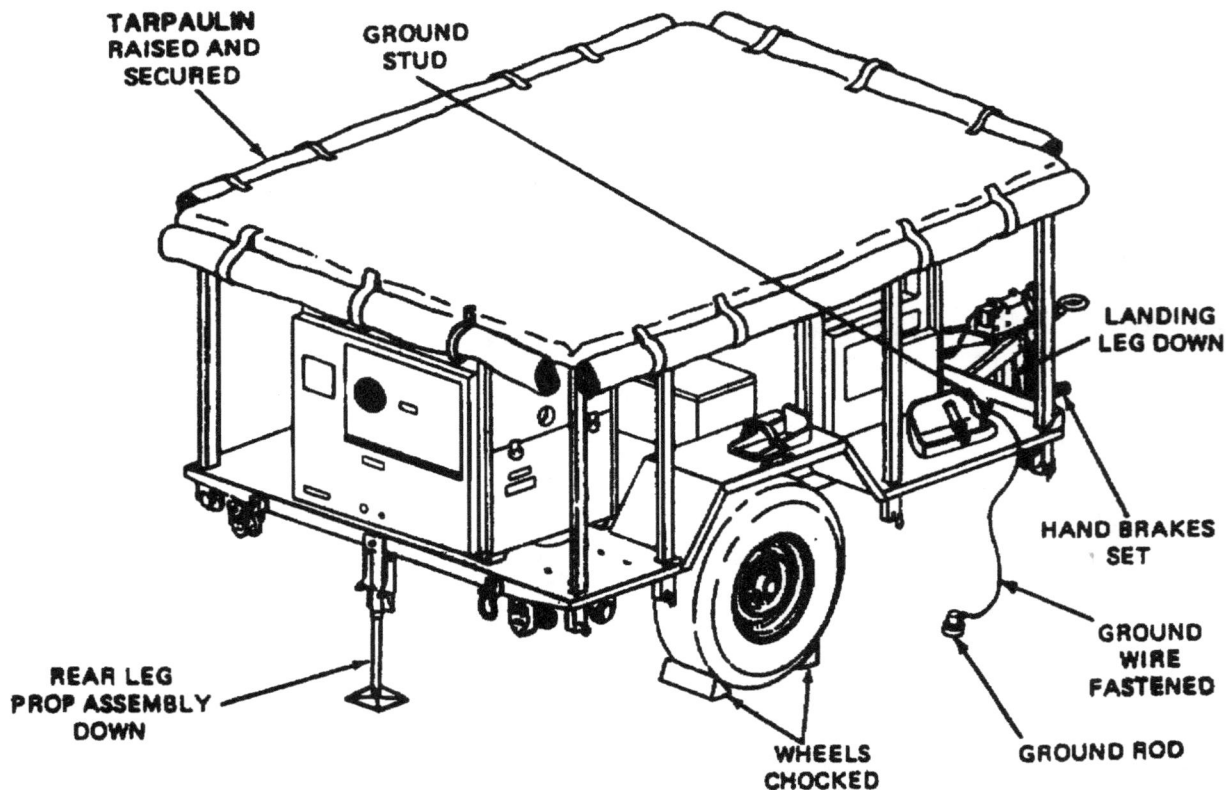

Figure 4-1. Installation of Power Plant; AN/MJQ-33 Shown.

WARNING

Do not operate generator sets until power plant is properly grounded. Serious injury or death by electrocution can result from operating an ungrounded power plant.

CAUTION

To avoid damage to equipment, make certain of voltage, frequency, and phase requirements of load being connected to generator set.

b. Grounding. Ensure generator sets are grounded to GROUND TERMINAL studs, two each, on trailer body. Using ground wire supplied, connect each of the power plant's two ground studs to a suitable ground as described below. The following sources of a good ground are listed in order of preference.

NOTE
As a substitute for the supplied ground wire, any copper wire of at least No. 6 AWG may be used.

   (1) Underground water system. Ground power plants to one of the accessible pipes in an underground water system. Make certain underground pipe is made of metal and there is no insulation, such as a water meter, between ground wire and earth.

   (2) Ground rod. Drive ground rod a minimum of 8 feet into earth. A ground rod must have a minimum diameter of 5/8-inch if solid, or 3/4-inch if pipe.

NOTE
It may be necessary to saturate the area around ground rod with water if soil conditions are dry.

c. External Fuel Line Connection. (See Figure 4-2). The power plant generator sets can be fueled from an external source such as a 5-gallon fuel can or 55-gallon drum. This eliminates the need for frequent refilling of each generator's fuel tank during long intervals of operation.

   (1) Remove fuel can adapter and fuel pickup tube from storage locations on power plants and assemble by threading pickup tube into adapter.

   (2) Thread one end of auxiliary fuel line onto fuel can adapter fitting and tighten.

   (3) Connect free end of auxiliary fuel line to AUXILIARY FUEL CONNECTION. This connection is located directly to the left of the fuel drain on the left-hand side of generator set.

   (4) Insert fuel can adapter in external fuel source and secure by pressing down on lever.

   (5) Set MASTER SWITCH on control panel to RUN AUX FUEL position.

4-7

NOTE
When generator set is run on auxiliary fuel, as described above, fuel is first pumped into generator set fuel tank by auxiliary fuel pump. Fuel is then fed to generator set engine from fuel tank.

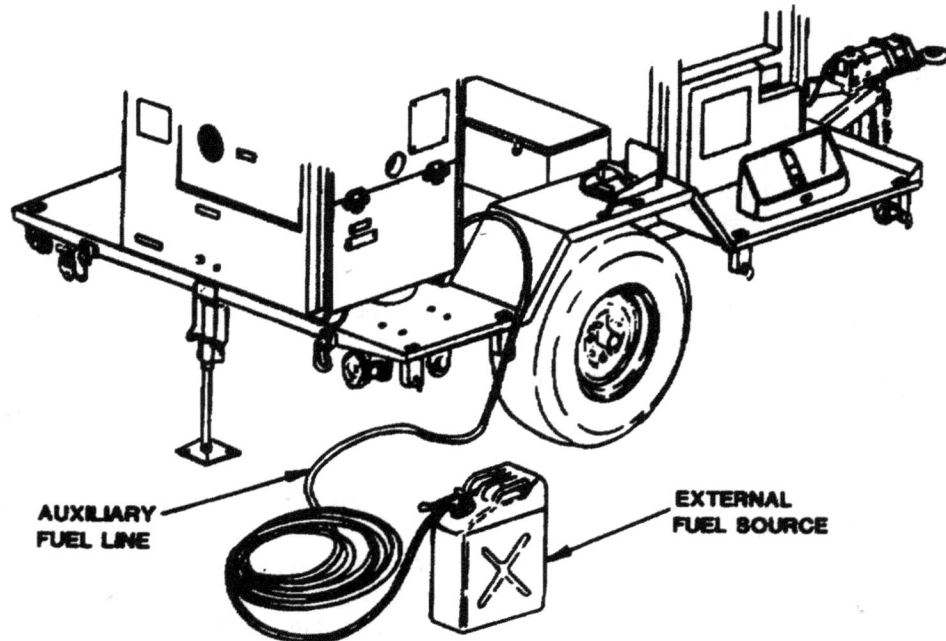

AUXILIARY FUEL LINE

EXTERNAL FUEL SOURCE

Figure 4-2. External Fuel Line Connection.

4-3. DISMANTLING FOR MOVEMENT.

Because the power plants are designed to be mobile, a minimum amount of effort is required to relocate at a new worksite. Procedures are as follows:

a. Disconnect power plants from system or equipment being powered.
b. Disconnect ground cables from source of ground and from power plants' GROUND TERMINAL studs. Roll up cable and store in accessory box for the AN/MJQ-33 or on rack assembly of AN//JQ-32.
c. Using slide hammer, remove ground rods. Disassemble, clean, and stow ground rods in accessory box for the AN/MJQ-33 or on rack assembly for the AN/MJQ-32.
d. Disconnect power plants from external fuel source, if applicable.
e. Stow any remaining authorized equipment for the AN/KIQ-33 in accessory box.

f. Stow any remaining authorized equipment for the AN/MJQ-32 on designated areas of the AN/MJQ-32.

g. Secure fire extinguisher and fuel cans in their respective mounting brackets.

h. Lower and secure tarpaulin in place on the AN/MJQ-33 power plant.

i. Place antenna and antenna accessories into stowage rack (AN/MJQ-32) and on mast mounts and supports. Secure in place.

j. Remove locking pin from leg prop assembly on rear of trailers. Swing leg prop back and up into traveling position and secure with pin.

k. Attach power plant to towing vehicle. (Refer to TM 9-2330-202-14&P).

l. Release trailer handbrakes.

4-4. REINSTALLATION AFTER MOVEMENT.

After movement to a new worksite, install power plants in accordance with paragraph 4-2.

Section II. REPAIR PARTS, SPECIAL TOOLS, SPECIAL TEST, MEASUREMENT AND DIAGNOSTIC EQUIPMENT (TMDE)

4-5. TOOLS AND EQUIPMENT.

There are no special tools or equipment required to maintain the AN/MJQ-32 and AN/MJQ-33 power plants.

4-6. MAINTENANCE REPAIR PARTS.

Repair parts for maintenance of these power plants are listed and illustrated in the repair parts and special tools list in Appendix C of this manual.

Section III. LUBRICATION INSTRUCTIONS

4-7. GENERAL.

Detailed instructions for the lubrication of the major components of the power plants are contained in the applicable Lubrication Orders (LO's). Refer to DA Pam 25-30 to ensure that the latest editions of the LO's are used. This section contains lubrication instructions that are not included in the Lubrication Orders.

4-9

4-8. GENERATOR LUBRICATION.
Refer to TM5-6115-615-12 for generator set Lubrication Order. Refer to items 2 thru 4, Appendix F, for lubricating items.

4-9. TRAILER ASSEMBLY LUBRICATION.

a. Trailer Lubrication. Refer to TM9-2330-202-14&P for trailer Lubrication Order. Refer to items 2 thru 6, Appendix F, for lubricating items.

b. Leg Prop Assembly Lubrication. The rear leg prop assembly is a modification to the standard M116A2 trailer and, as such, does not appear in the associated LO. Semiannually lubricate leg prop assembly as follows:

WARNING
Clean parts in a well-ventilated area. Avoid inhalation of solvent fumes and prolonged exposure of skin to cleaning solvent. Wash exposed skin thoroughly. Dry cleaning solvent, PD-68U, used to clean parts is potentially dangerous to personnel and property. Do not use near open flame or excessive heat. Flash point of solvent is 100°F to 138°F (38°C to 59°C).

(1) Clean lubrication fitting (1, figure 4-3) and area around lubrication points with PO-680 (item 1, Appendix F) or equivalent.
(2) Inject sufficient GAA (item 6, Appendix F) grease into hydraulic fitting to lubricate screw threads (2) inside leg base (3).

NOTE
Refer to Lubrication Order in TM9-2330-202-14&P for lubricating oils specified for use within different anticipated temperature ranges.

(3) Apply OE (items 2 or 3, Appendix F) lubricating oil to both ends of leg prop assembly pivot shaft (4).

Section IV. PREVENTIVE MAINTENANCE CHECKS AND SERVICES

NOTE
The PMCS chart in this section contains all necessary unit preventive maintenance checks and services for this equipment.

4-10

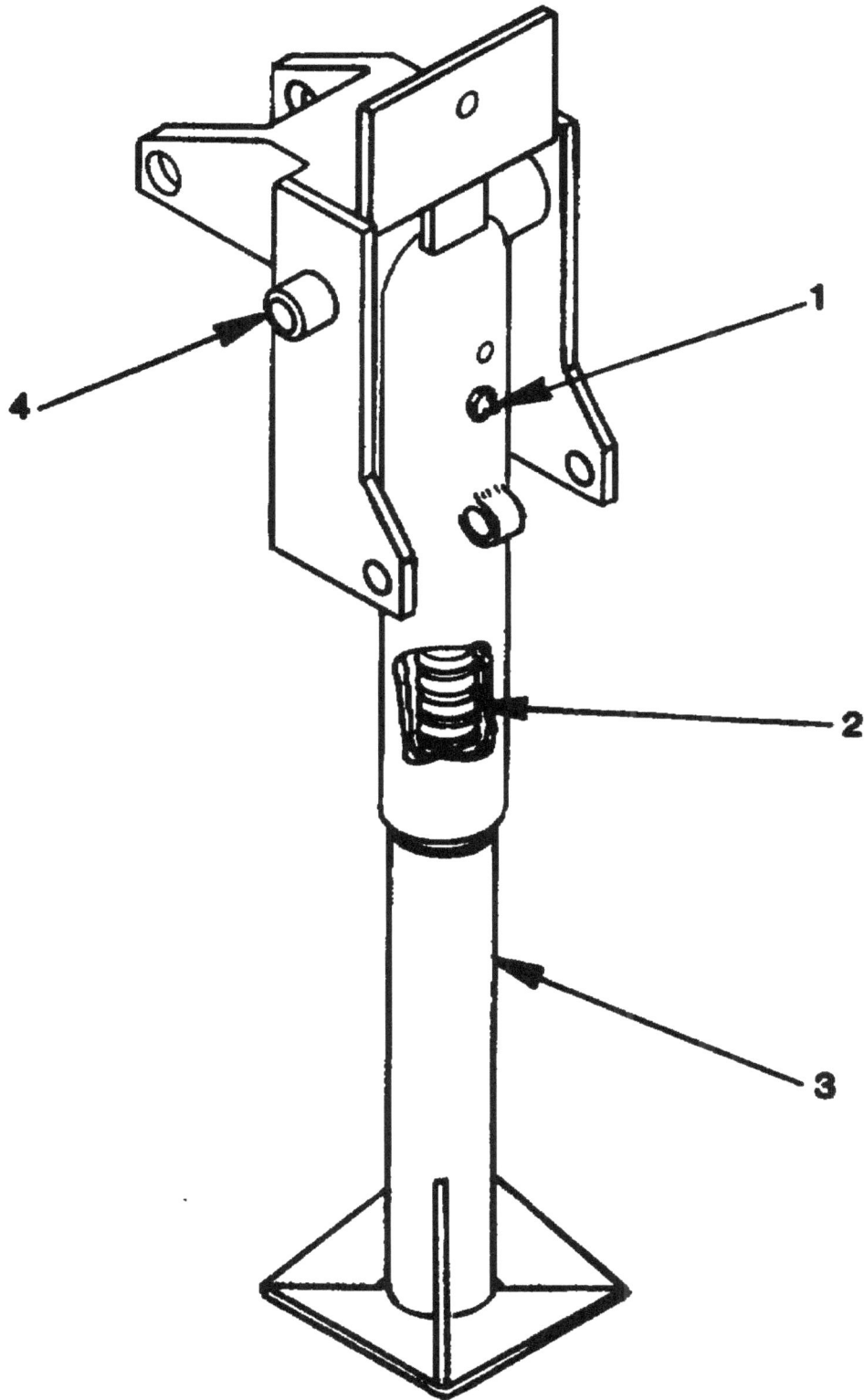

Figure 4-3. Rear Leg Prop Lubrication Points.

4-10. GENERAL.

The trailer assemblies and generator sets of the power plants must be inspected and service systematically to ensure that the power plants are ready for operation at all times. Inspection will allow defects to be discovered and corrected before they result in serious damage or failure. Table 4-1 contains a tabulated list of preventive maintenance checks and services to be performed by unit maintenance personnel. All of the unit PMCS on the trailers are scheduled to be performed semi-annually. Unit PMCS on the generator sets are scheduled weekly or on a per-hours-of-operation basis. The running time meter on the control panel is used to determine the generator set operating time. Using the following as a guide, do the checks and services at the intervals shown. Observe all CAUTIONS and WARNINGS.

    a. For PMCS performed on an operating time basis, perform your hourly (H) PMCS as close as possible to the time intervals indicated.

NOTE
For units in continuous operation, perform PMCS before starting operation if continuous operation will extend service interval past that which is shown.

    b. Perform your weekly (W) PMCS every week or 4U hours of generator set operating time.
    c. Perform your monthly (M) PMCS every month or 10u hours of generator set operating time.
    d. Do your semiannual (S) PMCS once every 6 months.
    e. If you discover a problem with the equipment, refer to Section VI, Troubleshooting. If you cannot correct the problem, refer to paragraph 4-12, Reporting Deficiencies.

4-11. EXPLANATION OF COLUMNS.
The following is a list of the PMCS table column headings with a description of the information found in each column.

    a. Item No. This column shows : the sequence in which to do the checks and services, and is used to identify the equipment area on the Equipment Inspection and Maintenance Worksheet, DA Form 2404.
    b. Interval. This column shows when each check is to be done.
    c. Item to be Inspected. This column identifies the general area or specific part where the check or service is to be done.
    d. Procedures. This column lists the checks or service you have to do and explains how to do them.

4-12. REPORTING DEFICIENCIES.

If you discover any problem with the equipment during PMCS that you are unable to correct, it must be reported. Refer to DA Pam 738-75u and report the deficiency using the proper forms.

Table 4-1. Unit Preventive Maintenance Checks and Services - Continued
H - Hours of operation W - Weekly M - Monthly S - Semiannually
(As indicated) (40 hours) (100 hours) (500 hours)

ITEM NO.	INTERVAL				Item to be inspected	Procedures
	H	W	M	S		
					WARNING Before performing any maintenance that requires climbing on or under-trailer, set trailer handbrakes, chock-wheels, and lower rear leg prop. Injury to personnel could result from trailer suddenly rolling or tipping. **NOTE** Generator set checks and services in this table are described as performed on a single generator set. These procedures must be performed on each of the two generator sets that make up the AN/MIQ-32 and AN/MJQ-33.	

4-13

Table 4-1. Unit Preventive Maintenance Checks and Services - Continued
H - Hours of operation W - Weekly M - Monthly S - Semiannually
(As indicated) (40 hours) (100 hours) (500 hours)

ITEM NO.	INTERVAL				Item to be inspected	Procedures
	H	W	M	S		
1				●	Generator Set (Remove panels according to TM5-6115-615-12)	Inspect generator set for fuel and oil leaks, loose or missing components andhardware, and unusual wearor deterioration. Cleangenerator set.
2				●	**NOTE** **Fuel system must be above freezing-temperature when draining waterand sediment from filters and tank.** Fuel filter	Open drains on fuel filter. Allow water and sedimentto drain into suitablecontainer until fuel runs clean.
3				●	Fuel Tank	Drain water and sediment (TM5-6115-615-12). Allowto drain into suitablecontainer until fuel runs clean.
4					Fuel Transfer Pumps Auxiliary Fuel Pumps	Clean or replace fuel filters of fuel pumps, asnecessary (TM5-6115-615-12). Change lubricating oil andfilter every 125 hours ofoperation (L05-6115-615-12).
5	125				Lubricating Oil and Filter	

Table 4-1. Unit Preventive Maintenance Checks and Services - Continued
H - Hours of operation W - Weekly M - Monthly S - Semiannually
(As indicated) (40 hours) (100 hours) (500 hours)

ITEM NO.	INTERVAL				Item to be inspected	Procedures
	H	W	M	S		
6	300				Batteries	Perform a hydrometer teston batteries every 300hours, or quarterly. Referto TM 5-6115-615-12 fortest procedures.
7	100				Dust Valves on Air Cleaner	Clean out dust valve on aircleaner assembly every 100operating hours (more fre-quently under unusual con-tions).
8	1000				Air Cleaner	Clean every 1000 operatinghours or as con-ditionsdictate. Replace aircleaner every 2000 opera-ting hours.
9				●	Taillights	Inspect for broken orcracked lenses or defec-tivebulbs. Replace if neces-sary.
10				●	Intervehicular Cable	Check for cuts, breaks,frayed wires, or dam-aged plug.
11				●	Drawbar Ring	Check security for mount-ing. Inspect ring forexcessive wear.
12				●	Safety Chains	Inspect for broken linksor missing chain(s).
13				●	Reflectors	Inspect for any cracked,broken, or missing reflec-tors. Replace if neces-sary.

Table 4-1. Unit Preventive Maintenance Checks and Services - Continued
H - Hours of operation W - Weekly M - Monthly S - Semiannually
(As indicated) (40 hours) (100 hours) (500 hours)

ITEM NO.	INTERVAL				Item to be inspected	Procedures
	H	W	M	S		
14				●	Data Plates and Markings	Make sure data plates and markings are legible and not missing.
15				●	Support Leg Assembly	Inspect brackets and leg for bent or broken parts.
16				●	Rear Leg Prop Assembly	Inspect bracket and leg prop for bent or broken parts.
17				●	Suspension Assemblies	a. Inspect shackles, bearings, pins, leafsprings and spring eyes for damage or broken parts. b. Inspect mounting brackets for cracks or loose or missing hardware. c. Inspect shock absorbers for damage or leaks.
18				●	Axle	a. Check for damaged axle tube. b. Check for loose or missing U-bolts or nuts.
19				●	Wheels and Tires	a. Check serviceability of tires as indicated in TM 9-2610-200-24. b. Inspect for loose or missing stud nuts. Tighten or replace as necessary.

Table 4-1. Unit Preventive Maintenance Checks and Services - Continued
H - Hours of operation W - Weekly M - Monthly S - Semiannually
(As indicated) (40 hours) (100 hours) (500 hours)

ITEM NO.	INTERVAL				Item to be inspected	Procedures
	H	W	M	S		
20				●	Brakes	Adjust brake. (Refer to TM 9-2330-202-14&P).
21				●	Wheel Bearings	Clean and repack. (Refer to TM 9-2330-202-14&P).
22				●	Hydraulic Brake Tubes and Hoses	Inspect for dents, cracks, loose connections, and leaks. Repair or replace as necessary.
23				●	Trailer - Road Test	Perform road test, paying special attention to items that were repaired or ad-justed.

Section V. TROUBLESHOOTING

4-13. TROUBLESHOOTING.

Troubleshooting procedures for components unique to each power plant's end item are given in paragraph 4-14. Trouble-shooting information for the individual generator sets and trailer is contained in their respective technical manuals refer-enced below.

a. Generator Set Troubleshooting. Refer to TM5-6115-615-12 for troubleshooting procedures applicable to the generator set.

b. Trailer Troubleshooting. Refer to TM9-2330-202-14&P for troubleshooting procedures applicable to the trailer.

4-17

4-14. POWER PLANT TROUBLESHOOTING.

Table 4-2 contains troubleshooting information for locating and correcting operating troubles which may develop in compo-nents unique to either power plant end item. Each malfunction is followed by a list of tests or inspections which will help determine probable cause and corrective actions to take. Perform the tests/inspections and corrective actions in the order listed. This manual cannot list all malfunctions that may occur nor all tests or inspections and corrective actions. If a mal-function is not listed or cannot be corrected by listed corrective actions, notify your supervisor. Troubleshooting information pertains to both power plants.

NOTE
Before you use the table, be sure you have performed your PMCS.

Table 4-2. Troubleshooting.

MALFUNCTION
      TEST OR INSPECTION CORREC-
            TIVE ACTION

---

1. POWER IS ABSENT AT SWITCH BOX LOAD TERMINAL(S) WHEN ONE GENERATOR SET IS OPERATING.

      Step 1. Check that circuit breaker on generator set control panel is set to ON position.

          If circuit breaker is in OFF position, set to ON position.

      Step 2. Check for power output at generator set load terminals.

          If power is absent at generator set load terminals, troubleshoot generator set. (Refer to TM 5-6115-615-12).

      Step 3. Inspect power cable connections inside switch box for looseness or broken wire terminals.

          Tighten loose connections. If any wire terminals are bro-ken, replace cable (paragraph 4-24b) or refer to higher level of maintenance for repair.

Table 4-2. Troubleshooting - Continued

MALFUNCTION
TEST OR INSPECTION CORREC-
TIVE ACTION

WARNING

Make sure generator sets are shut down before performing any continuity checks. Failure to follow this precaution may result in death by electrocution.

Step 4. Perform continuity check on associated generator set power cable (paragraph 4-19a(4)).

 If cable is defective, replace cable (paragraph 4-24b) or refer to higher level of maintenance for repair.

Step 5. Perform continuity check on switch (paragraph 4-19a(1)).

 If switch is defective, notify higher level of maintenance.

2. POWER IS ABSENT AT ONE OR MORE SWITCH BOX LOAD TERMINALS WHEN EITHER GENERATOR SET IS OPERATING.

Step 1. Check wire associated with non-functioning load terminal(s) for loose or broken terminals inside switch box.

 Tighten loose connection(s). If any wire terminals are broken, replace wire (paragraph 4-22b).

Step 2. Perform continuity check on wire from switch to non-functioning load terminal(s) (paragraph 4-19a(2)).

 Replace defective wire(s) (paragraph 4-22b).

Step 3. Perform continuity check on switch (paragraph 4-19a(1)).

 If switch is defective, notify higher level of maintenance.

4-19

Table 4-2. Troubleshooting - Continued

MALFUNCTION
    TEST OR INSPECTION CORREC-
        TIVE ACTION

3. **ONE OR BOTH INDICATOR LIGHTS FAIL TO LIGHT WHEN ASSOCIATED GENERATOR SET IS OPERATING AND ROTARY SWITCH IS SET TO ON POSITION FOR THAT GENERATOR.**

    Step 1. Check if bulb is defective.

        If bulb is defective, replace.

    Step 2. Check wires inside switch box associated with non-function-ing light for loose or broken wire terminals.

        Tighten loose connections. If any terminals are broken, replace wire(s) (paragraph 4-22b) or refer to higher level of maintenance for repair.

    Step 3. Perform continuity check on wires associated with non-functioning indicator (paragraph 4-19a(3)).

        If wires are defective, replace indicator light and wire assembly (paragraph 4-23b).

    Step 4. Perform continuity check on indicator housing (paragraph 4-19a(3)).

        If housing is defective, replace indicator light and wire assembly (paragraph 4-23b).

Section VI. RADIO INTERFERENCE SUPPRESSION

4-15. GENERAL METHODS USED TO ATTAIN PROPER SUPPRESSION.
Essentially, suppression is attained by providing a low resistance path to ground for stray currents. The methods used include shielding ignition and high-frequency wires, grounding the frame with bonding straps, and using filtering systems.

4-16. RADIO INTERFERENCE SUPPRESSION COMPONENTS.
All component parts of the power plant's end item, whose primary or secondary function is radio interference suppression, are on the generator sets. Refer to TM 5-6115-615-12 for location of radio interference suppression components.

Section VII. MAINTENANCE OF GENERATOR SETS

4-17. GENERAL.

For generator set maintenance procedures, refer to TM5-6115-615-12.

Section VIII. MAINTENANCE OF SWITCH BOX

4-18. GENERAL.

The switch box consists of a switch, post terminals wiring, light and wire
assembly, cables, and related hardware.

WARNING

Ensure generator sets are shut down before performing any maintenance on switch box. Failure to observe this precaution may result in injury or death by electrocution.

4-19. SWITCH BOX MAINTENANCE.

This Task covers 1. Test 3. Removal

2. Repair 4. Installation

Initial Setup:

1. Tools - Multimeter AN/PSM-45 (6625-01-139-2512)

- General Mechanics Tool Kit (5180-00-177-7033)

1. Materials/Parts - Post Terminals (96906); MS393347-2

- Electrical Leads (97403); 13212E3567-1/-2/-3/-4

- Light and Wire Assembly (97403); 13212E3560

- Cables (97403); 13212E3571-4/-5, 13212E3570-3/-4

4-21

a. Test. To isolate the source of an electrical system problem, perform continuity tests on the components of the switch box as described below. Refer to the schematic diagram inside the switch box (shown in figure 4-4) to locate and identify the test points indicated in these procedures. The switch boxes, except for power cable lengths, are identical for both the AN/MJQ-32 and the AN/MJQ-33 power plants.

(1) Switch Test.

(a) Remove 16 screws (1, figure 4-5), lockwashers (2), and flat washers (3) from switch box front panel (4) and pull panel forward.

(b) Set multimeter for continuity tests.

(c) Set rotary switch (5) to GEN 1 position.

(d) Touch one probe to switch terminal (6) AI and remaining probe to switch terminal L1. Repeat test between terminals A2 and L2 and terminals A3 and L3. If multimeter does not indicate continuity exists between each pair of terminals, switch is defective. Notify higher level of maintenance.

(e) Set rotary switch (5) to GEN 2 position and repeat step (d) above, substituting terminals B1, B2, and B3 for A1, A2, and A3.

(2) Electrical Leads (Wiring Test).

(a) Remove 16 screws (1, figure 4-5), lockwashers (2), and flat washers (3) from switch box front panel (4) and pull panel forward.

(b) Set multimeter for continuity tests.

(c) Test each wire (7) between its switch terminal (6) and load (post) terminal (8) by touching probes to each of following pair of test points: L1 on switch to load (post) terminal L1, L2 on switch to load (post) terminal L2, and L3 on switch to load (post) terminal L3. If multimeter does not indicate continuity between each pair of test points replace associated wiring.

(d) Test ground wire (10) by touching one probe to load terminal (8) LO on switch box terminal board and touch remaining probe to E2 (AC GROUND) (9) on switch box. If multimeter does not indicate continuity, replace associated wire.

4-22

WIRING DIAGRAM, SWITCH BOX, P/N 13205E5079

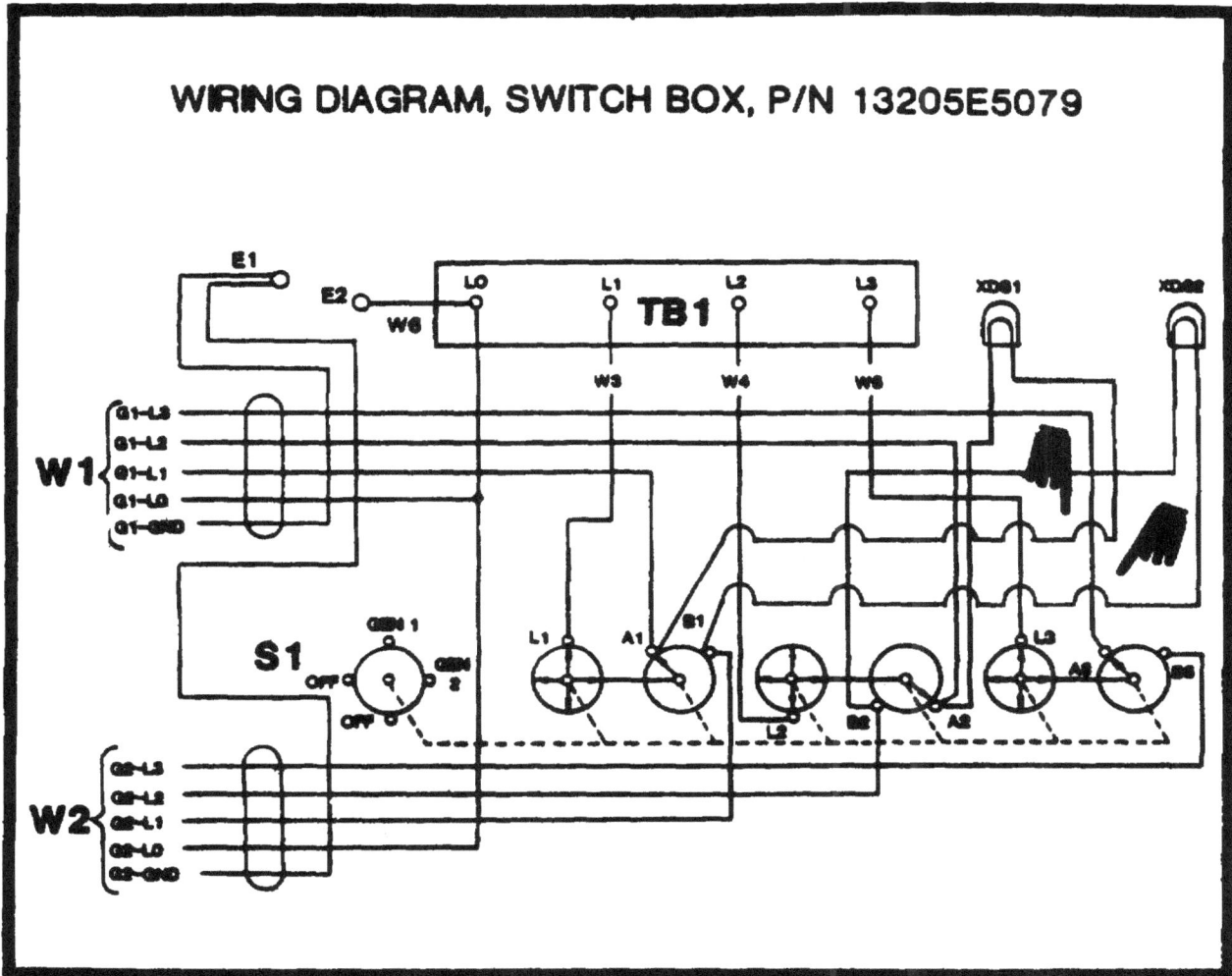

Figure 4-4. Five-Wire Switch Box Schematic Diagram. Diagram.

4-23

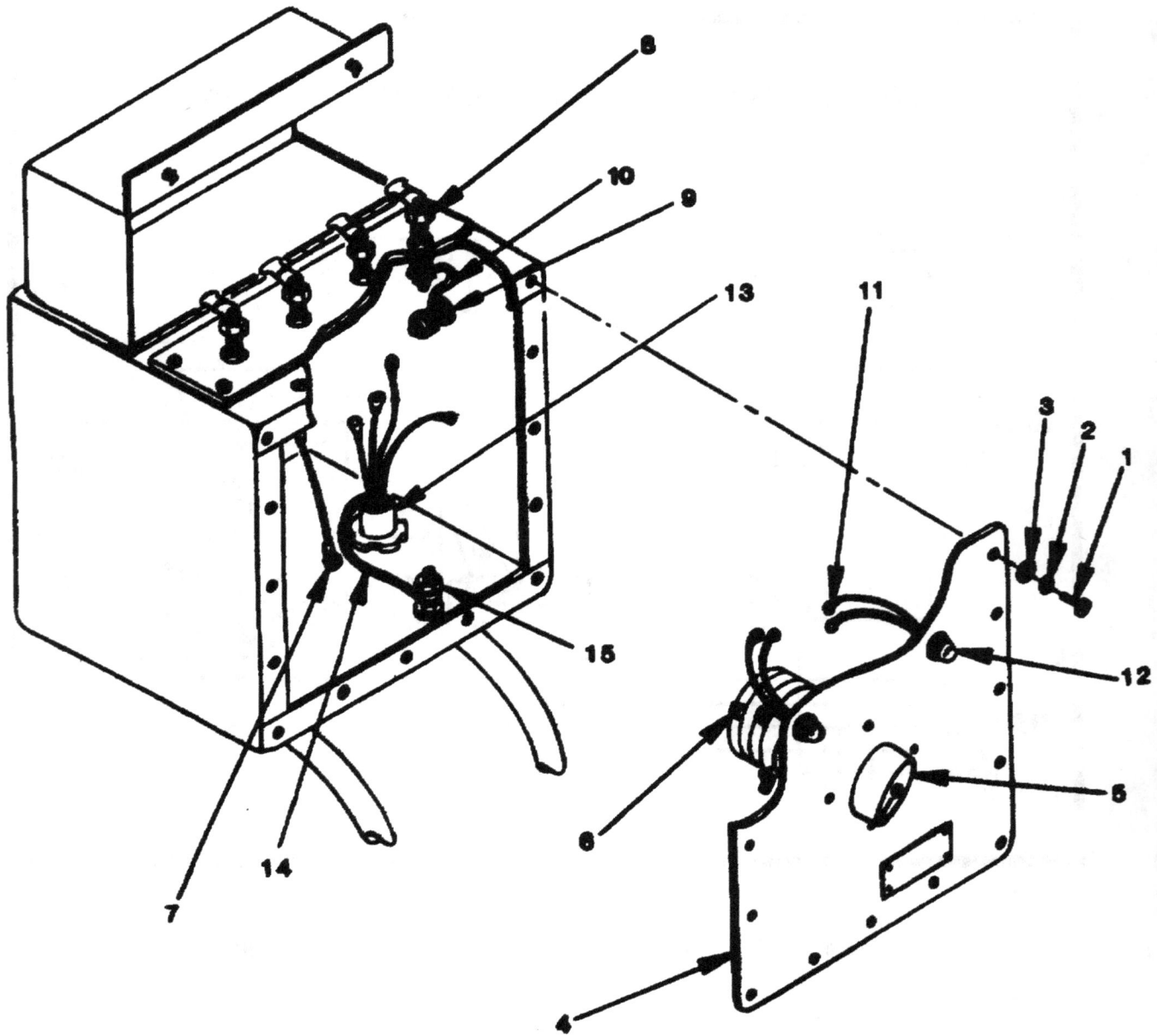

Figure 4-5. Switch Box Testing.

(3) Light and Wire Assembly Test.

    (a) Remove 16 screws (1, figure 4-5), lockwashers (2), and flat washers (3) from switch box panel (4) and pull panel forward.

    (b) Set multimeter for continuity test.

    (c) Test wires (11) and socket (12) associated with GEN 1 indicator light by testing for continuity between switch terminal (6) A2 and socket XDS1 and switch terminal A1 and socket XDS1. If multimeter does not indicate continuity exists on both of these wires, replace light and wire assembly (paragraph 4-23b).

    (d) Test wires and socket associated with GEN 2 indicator light by testing for continuity between switch terminal B2 and XDS2 socket, and switch terminal B1 and socket XDS1. If multimeter does not indicate continuity exists on both of these wires, replace light and wire assembly (paragraph 4-23b).

(4) Cable Assembly Test.

    (a) Remove 16 screws (I, figure 4-5), lockwashers (2), and flat washers (3) from switch box front panel (4) and pull panel forward.

    (b) Set multimeter for continuity testing.

    (c) Set switch box rotary switch (5) to GEN 1 position and locate power cable assembly (13) associated with switch position.

    (d) Test white wire by touching one probe of multimeter to load terminal LO on generator set and touching other probe to load terminal (8), associated with white wire, on switch box. Multimeter must indicate continuity between these points.

    (e) Repeat step (3) for black wire (between L1 on generator set and AI on switch), red wire (between L2 on generator set and A2 on switch), and blue wire (between L3 on generator set and A3 on switch).

    (f) Test the ground wire (14) by disconnecting the ground wire that runs from generator to skid mount ground stud; disconnect wire at skid mount. Test for continuity between free end of wire and ground stud EI, EQUIPMENT (FRAME) GROUND (15), on switch box.

Change 2 4-25

(g) If multimeter does not indicate continuity exists on each wire in the cable, replace power cable assembly (paragraph 4-24b).

(h) To test power cable assembly associated with GEN 2 switch position, perform steps (2) thru (6) substituting switch terminals B1, B2, and B3 for terminals A1, A2, and A3, at test points. Generator set load terminal designations and wire color-coding are identical for both power cables.

b. Repair. Repair of the switch box is accomplished by replacement of the post terminals, wiring, cables, and replacement or repair of the light and wire assembly.

c. Removal. (See figure 4-6). The switch boxes, one on each power plant, are located on the roadside rear fender extension for AN/MJQ-32 and on the roadside fender for AN/MJQ-33. Switch boxes, except for power cable lengths, are identical for both power plants.

WARNING
Make sure that generator set circuit breakers are in the OFF position before performing removal procedures on switchbox. Failure to follow this precaution may result in injury or death by electrocution.

(1) Removal.

(a) Remove wingnut (1, figure 4-6), lockwasher (2), and flatwasher (3) from EQUIPMENT (Frame) GROUND stud (4). Slide ground wire (5) off stud and remove remaining flat washer (3), nut (6), and star washer (7).

(b) Detach load cables (8) from load terminals on generator sets. (Refer to TM 5-6115-615-12.) (c) Remove four screws (9), lockwashers (10), and flat washers (11) securing switch box (12) to bracket (13).

(d) Remove switch box from bracket and pull cabling through cable brackets.

d. Installation.

(1) Position switch box (12) on bracket (13).

(2) Insert four flat washers (11), lockwashers (10), and screws (9) through bracket and into switch box. Tighten hardware to secure switch box.

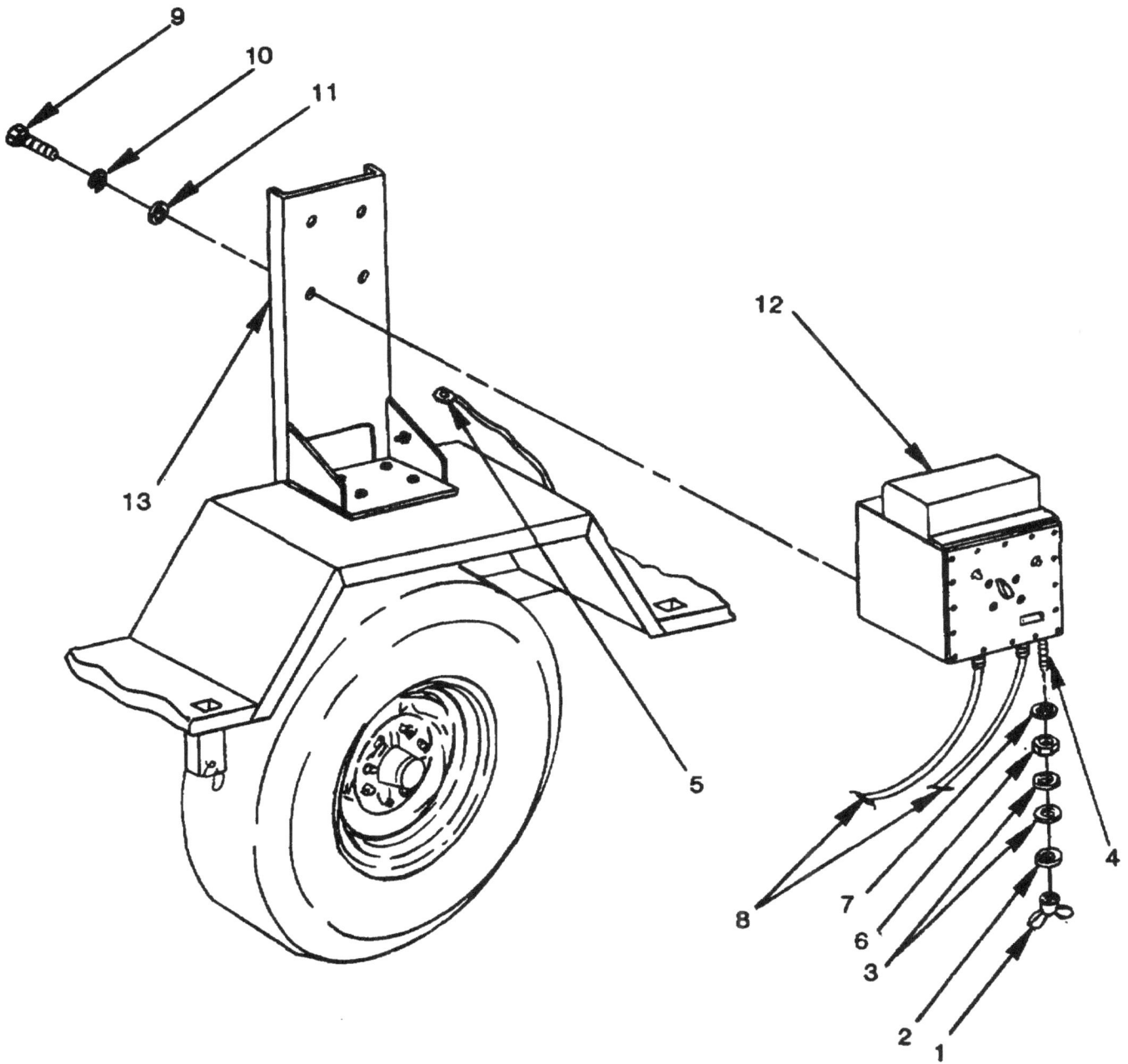

Figure 4-6. Switch Box Replacement.

(3) Install washer (7), nut (6), (tighten), flat washer (3), ground wire (5), flat washer (3), lockwasher (2) and wingnut (1). Tighten
    wingnut.
(4) Route cabling through cable brackets.
(5) Attach cables to generator set (refer to TM5-6115-615-12).

Section IX. SWITCH BOX SWITCH MAINTENANCE

4-20. SWITCH BOX SWITCH MAINTENANCE.

This Task covers Testing

Initial Setup:

1. Tools - General Mechanics Tool Kit (5180-00-177-7033)

   - Multimeter AN/PSM-45 (6625-01-139-2512)

1. Materials/Parts - None

WARNING
Ensure generator sets are shut down before performing any Maintenance on switch. Failure to follow this precaution may result in injury or death by electrocution.

Test. Refer to paragraph 4-19a(1).

Section X. POST (LOAD) TERMINAL/TERMINAL BOARD MAINTENANCE

4-21. POST (LOAD) TERMINAL/TERMINAL BOARD MAINTENANCE.

This Task covers 1. Removal

                                        2. Installation

Initial Setup:

1. Tools - General Mechanics Tool Kit (5180-00-177-7033)

    - Multimeter AN/PSM-45 (6625-01-139-2512)

1. Materials/Parts - Post Terminals (96906); MS39347-2 Termi-

    nal Board (97403); 13212E3560

WARNING
Make sure generator sets are shut down before performing any maintenance on terminals or terminal boards. Failure to follow this precaution may result in injury or death by electrocution.

There are four load terminals on the switch box terminal board. This procedure is typical for all four. When reconnecting wires, refer to the schematic inside the switch box (figure 4-4) and to the identification bands on the wires.

a. Removal

(1) Remove 16 screws (1, figure 4-7), lockwashers (2), and flat washers (3) securing cover (4) to switch box (5) and take cover off switch box.

NOTE
Before disconnecting wires, make sure wires are tagged to identify load terminal to which they attach.

(2) Working inside switch box (5), remove nut (6) and lockwasher (7) from terminal (8). Detach wire(s) (9) by sliding terminal lug(s) off stud.

4-29

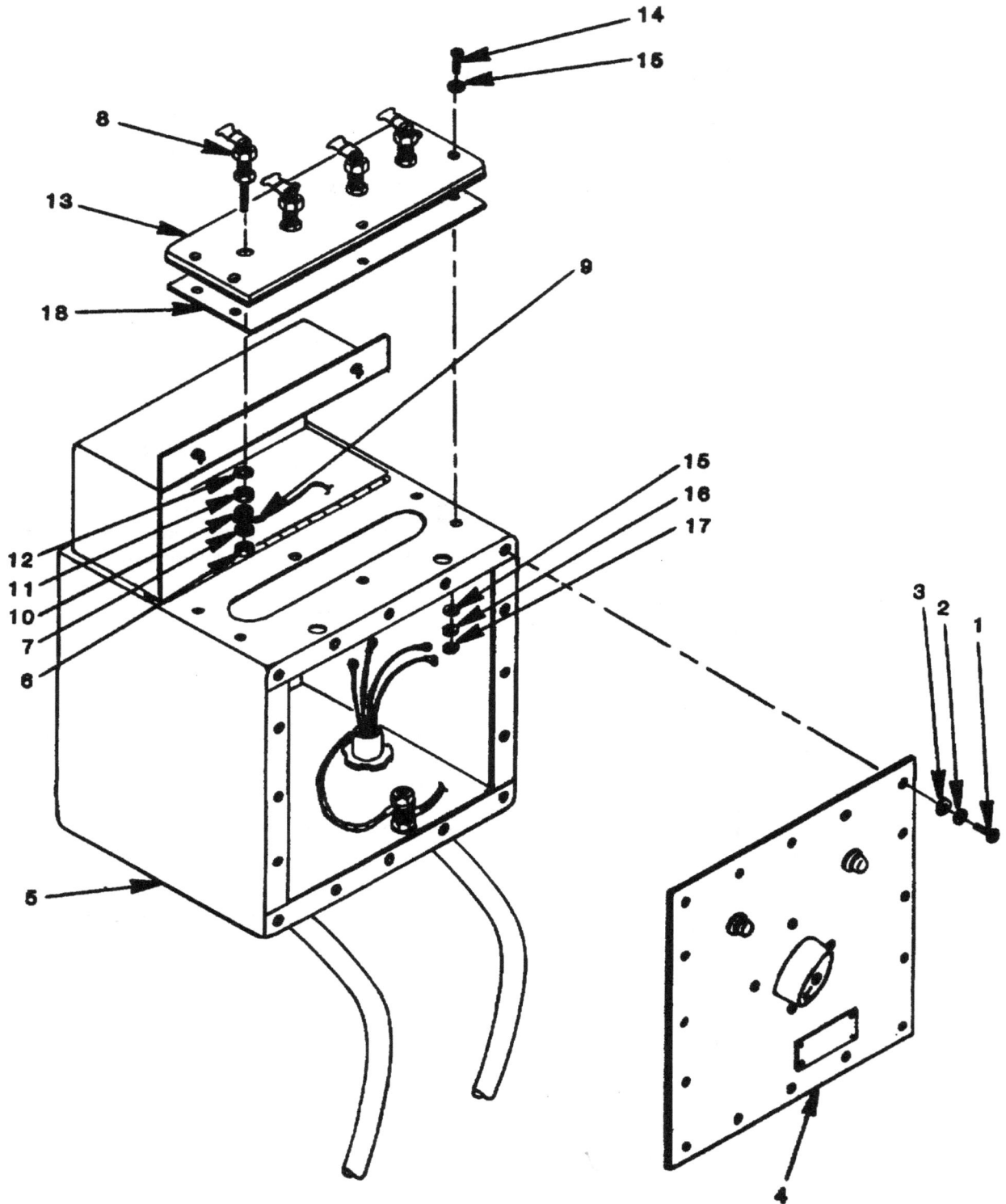

Figure 4-7. Load Terminal/Terminal Board Replacement.

(3) Remove lockwasher (10), nut (11) and internal-tooth lockwasher (12) and pull terminal (8) off terminal board (13).

(4) Remove 6 nuts (17), 6 lockwashers (16), 12 washers (15), and 6 screws (14) and take terminal board (13) and gasket (18) off switch box (5).

b. Installation.

(1) Position gasket (18) and terminal board (13) on switch box (5) and fasten with 6 screws (14), 12 washers (15), 6 lockwashers (16), and 6 nuts (17).

(2) Insert alignment tip of terminal stud (8) through small hole of terminal board (13) and into switch box (5).

(3) Install internal-tooth lockwasher (12) and nut (11) on terminal stud and tighten against inside of terminal board.

NOTE
Observe identification tags when installing internal wires.

(4) Slide lockwasher (10) and terminal lug(s) of wire(s) (9) on terminal stud (8).

(5) Install lockwasher (7) and nut (6) on terminal stud and tighten.

(6) Position cover (4) on switch box (5) and secure with 16 screws (1), lockwashers (2) and flat washers (3).

Section XI. ELECTRICAL LEADS (WIRING) MAINTENANCE

4-22. WIRING MAINTENANCE.

This Task covers 1. Test 3. Installation

2. Removal

Initial Setup:

1. Tools - Multimeter AN/PSM-45 (6625-01-139-2512)

   - General Mechanics Tool Kit (5180-00-177-7033)

1. Materials/Parts - Electrical Leads (97403); 13212E3567-1/-2/-3/-4

WARNING
Ensure generator sets are shut down before performing any maintenance on terminals or terminal boards. Failure to follow this precaution may result in injury or death by electrocution.

a. Test. Refer to paragraph 4-19a(2).

b. Removal. (See figure 4-8.) There are three wires connecting the switch to the load terminals; and one ground wire connecting load terminal LO to AC GROUND E2 (located on side of switch box). When attaching wires, refer to the schematic inside the switch box (figure 4-4) and to the identification bands on the wires.

(1) Removal (Switch-to-Load Terminal Wiring).

(a) Remove 16 screws (1, figure 4-8), lockwashers (2) and flat washers (3) securing cover (4) to switch box (5) and take cover off switch box.

(b) Remove nut (6) two lockwashers (7) and wire terminal lug (9) from terminal stud (8) inside switch box (5).

NOTE
When removing wires, tag switch terminals for identification.

(c) Remove screw (10), securing wire terminal lug (11) to threaded switch terminal (12). Remove wire from switch terminal.

(2) Installation (Switch-to-Load Terminal Wiring).

NOTE
Observe identification tags on switch terminal when installing wires.

(a) Position wire terminal (11) against underside of switch terminal (12). Insert screw (10) into wire terminal and tighten on switch terminal.

4-32

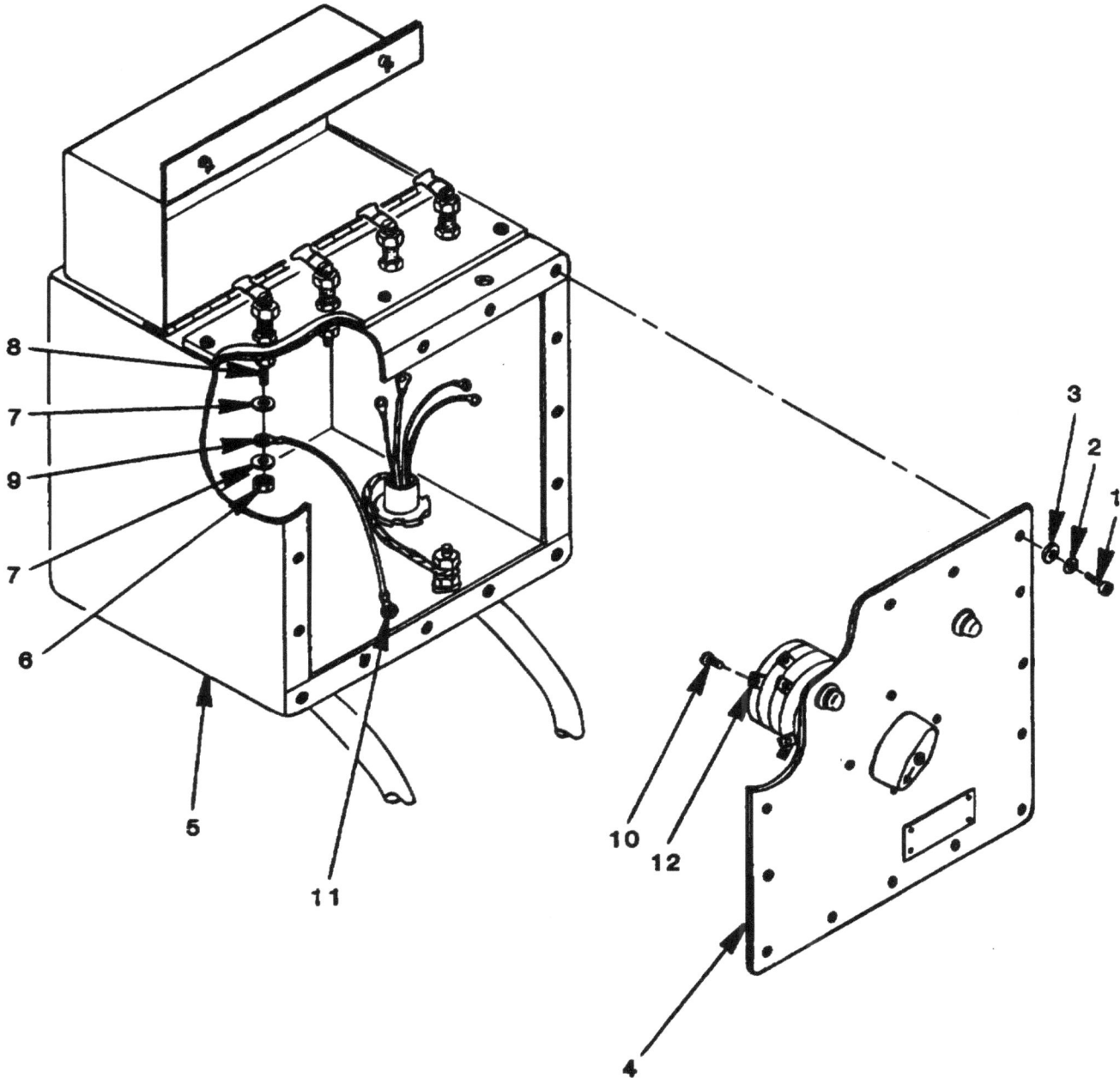

Figure 4-8. Switch-to-Load Terminal Wire Replacement.

(b) Slide one lockwasher (7) and terminal lug (9) on load terminal stud (8) and secure with lockwasher (7) and nut (6).

(c) Position cover (4) on switch box (5) and secure with 16 flat washers (3), lockwashers (2), and screws (1).

(3) Removal (Ground Wire).

(a) Remove 16 screws (1, figure 4-9), lockwashers (2) and flat washers (3) securing cover (4) to switch box (5) and take cover off switch box.

(b) Remove nut (6) and lockwasher (7) from load terminal (8) LO inside switch box.

(c) Slide power cable ground (white) wire terminals (9) and switch box ground wire (10) off load terminal (8) LO.

(d) Remove remaining lockwasher (7) from load terminal.

(e) Remove nut (11) and lockwasher (12) from AC GROUND terminal E2 (13) inside box. Remove ground wire (10).

c. Installation (Ground Wire).

(1) Install switch box ground wire (10) on AC GROUND terminal E2 (13) and secure with lockwasher (12) and nut (11).

(2) Install remaining end of ground wire (10) on load terminal (8) LO. Install power cable (white) wire terminals (9) on terminal LO and secure the three wires with lockwasher (7) and nut (6).

(3) Position cover (4) on switch box (5) and secure with 16 flat washers (3), lockwashers (2) and screws (1).

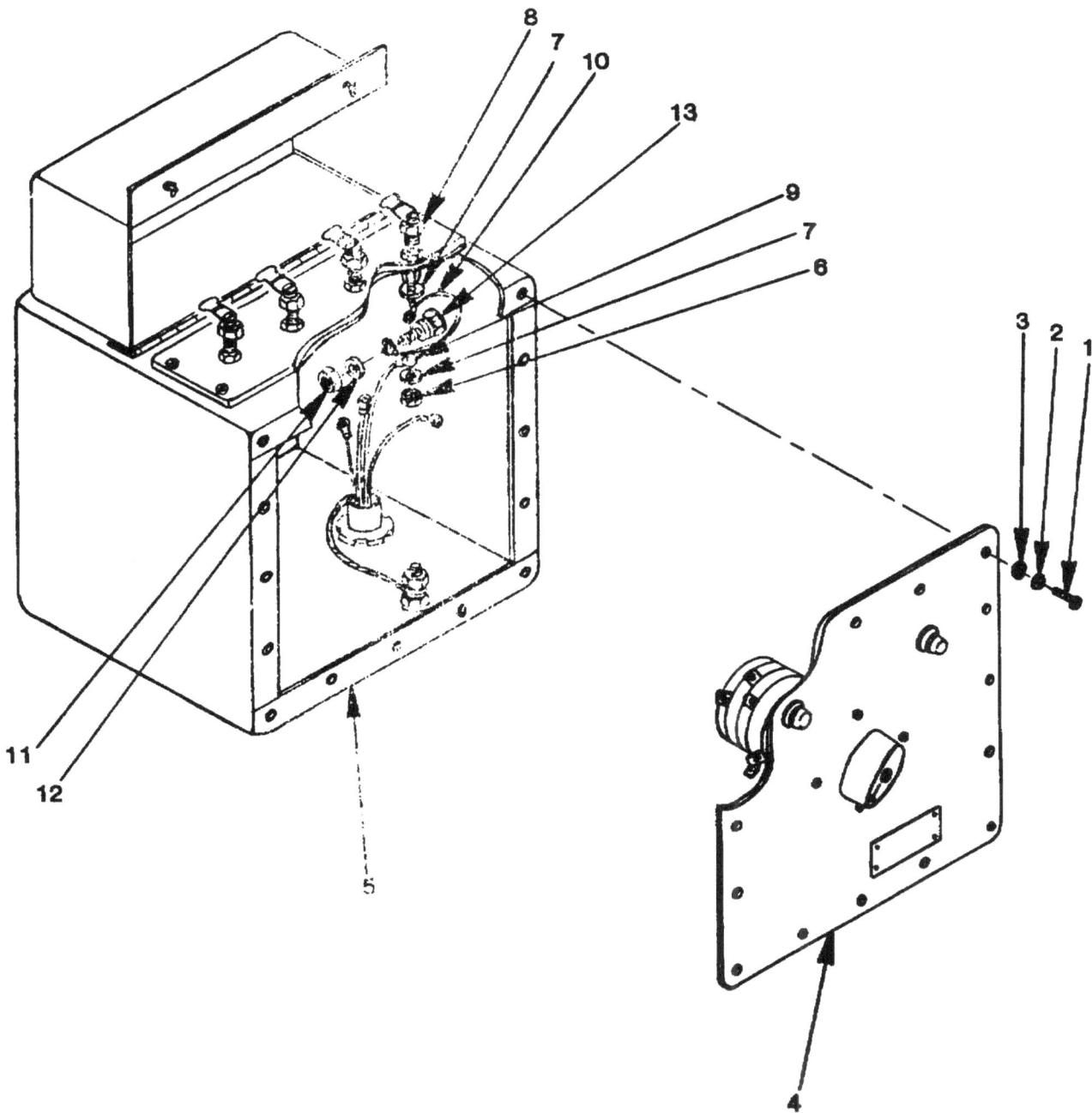

Figure 4-9. Switch Box Ground Wire Replacement.

Section XII. LIGHT AND WIRE ASSEMBLY MAINTENANCE

4-23. LIGHT AND WIRE ASSEMBLY MAINTENANCE.

This task covers: 1. Test 3. Repair

2. Removal 4. Installation

Initial Setup:
1. Tools - Multimeter AW/PSH-45 (6625-01-139-2512)
   - General Mechanics Tool Kit (5180-00-177-7033)
   - Soldering Gun GTTA-3 (3439-00-004-0915)
1. Materials/Parts - Lens (72619); 181-0937-003
   - Light (58224); NE2G
   - Housing (72619); 181-8836-09-553
   - Light and Wire Assembly (97403); 13212E3560

WARNING
Make sure generator sets are shut down before performing any maintenance on wiring. Failure to follow this precaution may result in injury or death by electrocution.

a. Test. Refer to paragraph 4-19a(3).

b. Removal (See figure 4-10). There are two indicator light and wire assemblies on the switch box. This procedure is typical for both. When attaching wires, refer to the schematic inside the switch box (figure 4-4).

(1) Remove 16 screws (1, figure 4-10), lockwashers (2) and flat washers (3) securing cover (4) to switch box (5) and take cover off switch box.

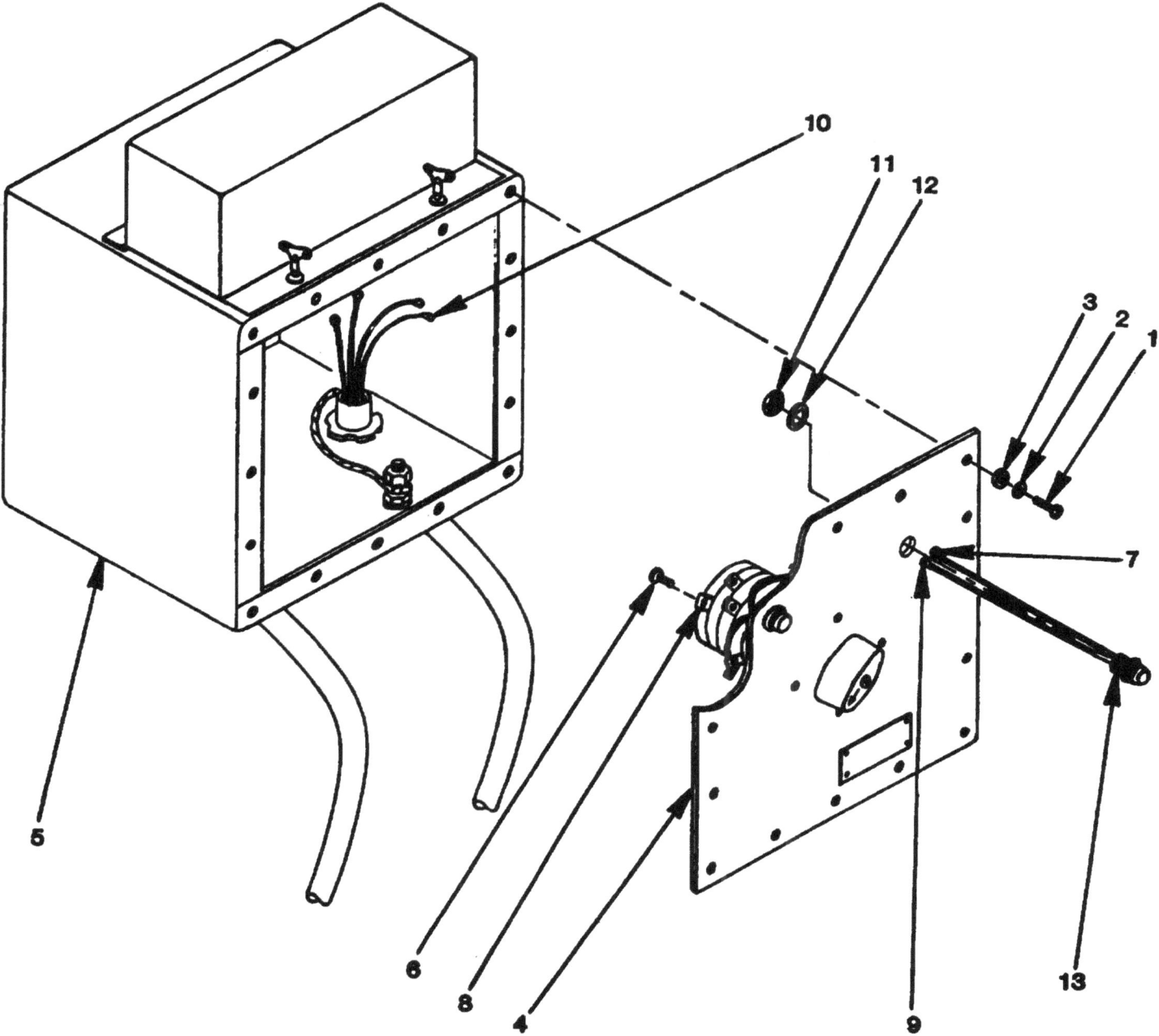

Figure 4-10. Switch Box Light and Wire Assembly Replacement.

(2) Remove screw (6) attaching each indicator light wire terminal (7) to its respective threaded switch terminal (8). Tag and disconnect power cable wire (10) and indicator light wire (9) from switch terminal.

(3) Unscrew nut (11) and lockwasher (12) securing indicator light housing (13) to cover (4) and slide nut and lockwasher off wires.

(4) Pull out light and wire assembly through cover.

c. Repair. The light and wire assembly is repaired by replacing defective lens, bulbs, housing, or soldering broken wires. Soldering shall be done in accordance with TB SIG 22.

d. Installation.

(1) Feed wires through hole in cover (4) and fit light housing (13) into hole.

(2) Slide lockwasher (12) and nut (11) over both indicator light wires (9) and tighten.

(3) Position terminal (7) of each indicator light wire (9) against underside of switch terminal (8). Position terminal of load cable wire (10) against terminal of indicator light wire. Insert screw (6) through wire terminals and tighten into threaded switch terminal (8).

(4) Position cover (4) on switch box (5) and secure with 16 flat washers (3), lockwashers (2), and screws (1).

Section XIII. POWER CABLE ASSEMBLY MAINTENANCE

4-24. POWER CABLE ASSEMBLY MAINTENANCE.

This task covers: 1. Test 3. Installation
2. Removal

Initial Setup:
1. Tools - Multimeter AN/PSM-45 (6625-01-139-2512)
   - General Mechanics Tool Kit (5180-00-177-7033)
1. Materials/Parts - Cable Assembly (94703); 13212E3571-4/-5,
   13212E3570-3/-4

WARNING
Make sure generator sets are shut down before performing any maintenance on wiring. Failure to follow this precaution may result in injury or death by electrocution.

a. Test. Refer to paragraph 4-19a(4).

b. Removal (See figure 4-11). There are two power cable assemblies on each of the switch boxes on the power plants - one power cable assembly for each generator set. This procedure is typical for both. When attaching wires, refer to the schematic inside the switch box (figure 4-4) and to the identification bands on each of the five wires that make up each power cable.

(1) Removal.

(a) Disconnect switch box power cable from load terminals on generator set. Refer to TM5-6115-615-12.

(b) Remove 16 screws (1, figure 4-11), lockwashers (2) and flat washers (3) securing cover (4) to switch box (5), and take cover off switch box.

NOTE
Tag switch terminals for identification when removing wires.
If identification bands on wires are illegible, tag wires.

(c) Remove screws (6) attaching each of three wires (7) (black, red, and blue) to its respective switch terminal (8).
Remove wires.

(d) Remove nut (9) and lockwasher (10) from load terminal (13) LO inside switch box (5). Take off white wire (11) associated with power cable being removed along with lockwasher (12).

(e) Remove nut (14), lockwasher (15), and flat washer (16) from EI (18). Take off green wiring (17) and other green wiring (associated with power cable being removed), if necessary.

(f) Loosen power cable clamping nut (19) on outside of switch box (5) and pull cable through cable clamp and out of box.

(g) Pull power cable through cable brackets and remove power cable from trailer.

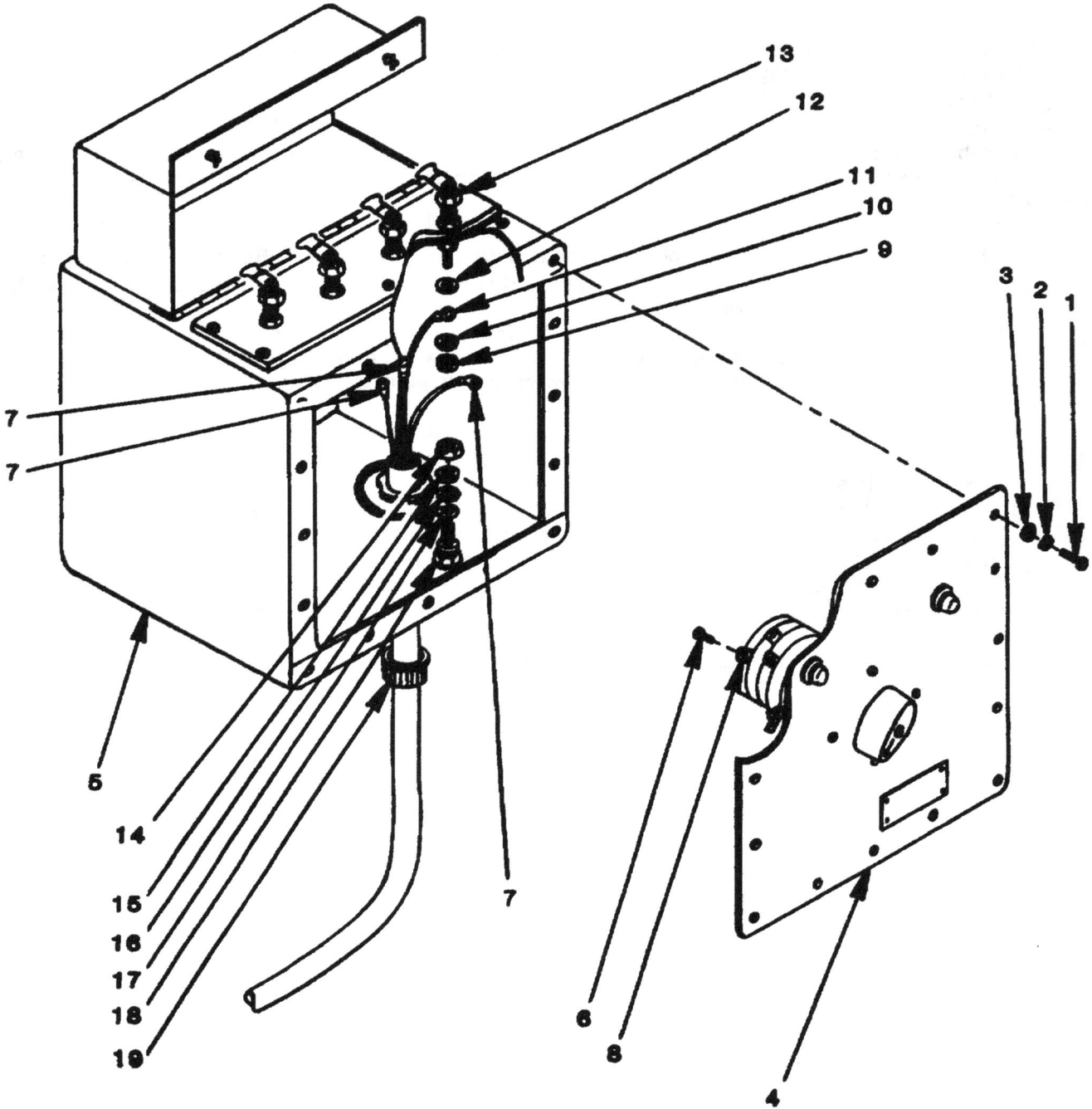

Figure 4-11. Switch Box Power Cable Replacement.

c. Installation.

(1) Route power cable through power cable brackets.

(2) Feed cable through cable clamping nut (19) and through hole into switch box (5).

(3) Tighten clamping nut (19) on outside of switch box (5) to prevent cable from moving.

(4) Slide washer (12) and terminal of white wire (11) on load terminal (13) LO and secure with lockwasher (10) and nut (9).

(5) Slide terminal of green wiring (17), and other green wiring, if necessary, on E1 (18) and secure with flat washer (16), lockwasher (15), and nut (14).

NOTE
Observe identification tags when installing wires.

(6) Position each of three wires (7) (black, red and blue) against underside of its respective switch terminal (8).

Insert screw (6) through wire terminal and tighten on threaded switch terminal. Where an indicator light wire is also mounted to the switch terminal, the indicator light wire shall be positioned against the underside of the switch terminal and the power cable wire shall be positioned against the indicator light wire.

(7) Position cover (4) on switch box (5) and secure with 16 screws (1), lockwashers (2) and flat washers (3).

(8) Connect power cable to load terminals on generator set. (Refer to TM5-6115-615-12).

Section XIV. ACCESSORY BOX MAINTENANCE

4-25. ACCESSORY BOX MAINTENANCE, AN/MJQ-33 ONLY.

This task covers: 1. Removal

2. Installation

Initial Setup:
1. Tools - General Mechanics Tool Kit (5180-00-177-7033)
2. Materials/Parts - Accessory Box (97403); 13226E7737

Power plant, AN/MJQ-33 only, is equipped with an accessory box which is mounted to the trailer bed between the front and rear generator sets.

a. Removal.

(1) Remove, two each, fuel can brackets (paragraph 4-27) situated in front of accessory box.

(2) Remove four screws (1, figure 4-12), eight flat washers (2), and four nuts (3) securing accessory box (4) to trailer bed (5).

(3) Remove accessory box from trailer bed.

b. Installation.

(1) Position accessory box (4) on trailer bed (5).

(2) Insert four screws (1) with flat washers (2) through accessory box mounting brackets and trailer bed (5).

(3) Working under trailer, install one flat washer (2) and nut (3) on each screw (1). Tighten hardware to secure accessory box (4).

(4) Install fuel can brackets (paragraph 4-27).

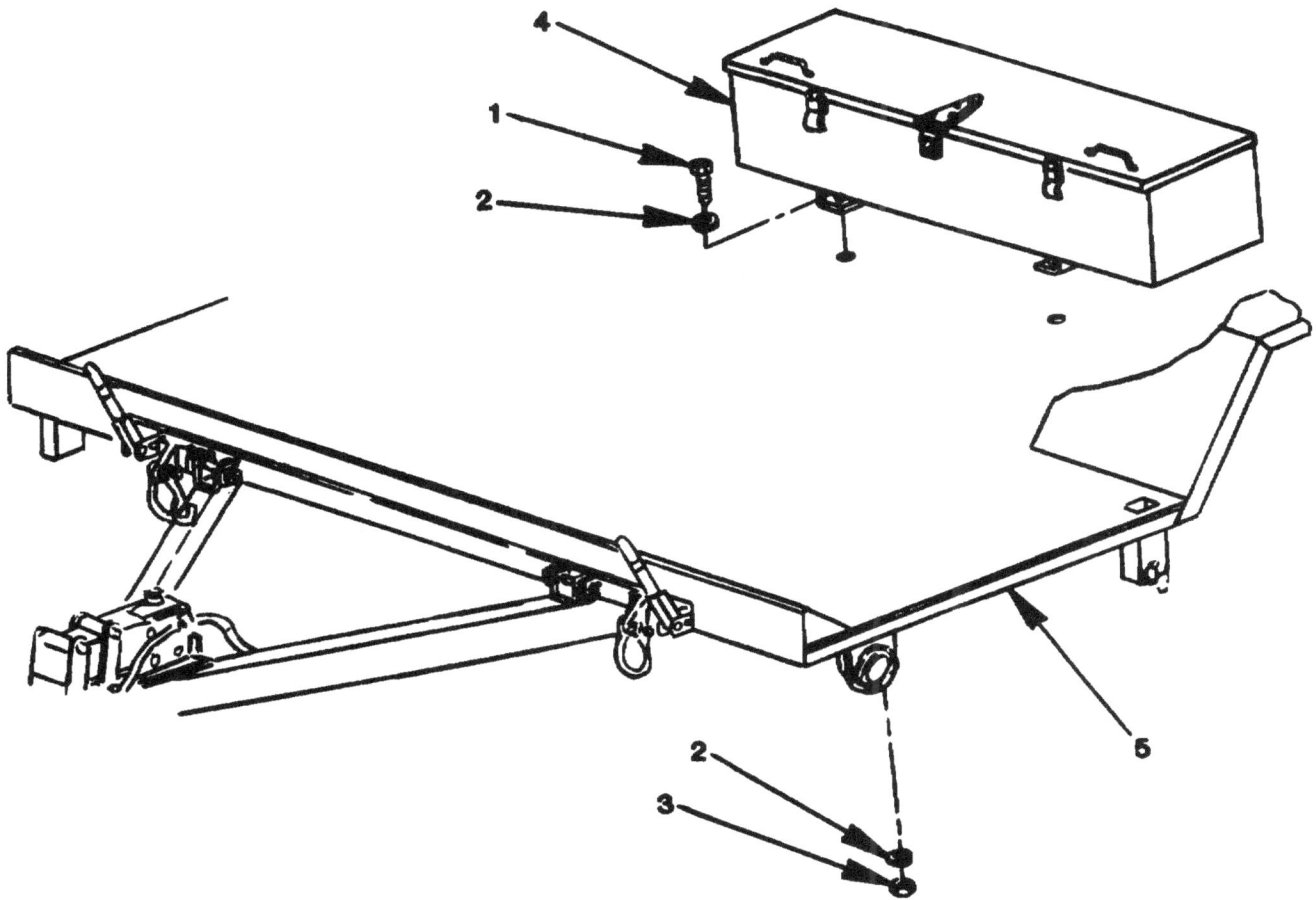

Figure 4-12. Accessory Box Replacement on AN/MJQ-33.

Section XV. STOWAGE RACK MAINTENANCE

4-26. STOWAGE RACK MAINTENANCE, AN/MJQ-32 ONLY.

This task covers: 1. Repair 3. Installation

                        2. Removal

Initial Setup:
    1. Tools - General Mechanics Tool Kit (5180-00-177-7033)
        - Drill, 1/4-inch (5130-00-807-3009)
    2. Materials/Parts - Rack Assembly (97403); 13228E9902
                  - Clamps (97403); 13205E5137-2
                  - Runners (97403); 13205E5120, 13205E5121,
13025E5123

                  - Strap Fastener (97403); 13218E5091

---

    a. Repair. (See figure 4-13). Repair of the rack assembly is limited to replacement of clamps, runners, and strap fasteners. If required, repaint in accordance with MIL-T-704 and MIL-C-46168. Replacement is as follows:
(1) Clamps.
    (a) Remove two nuts (6, figure 4-13), flat washers (7), and screws (8) from either one of two each leaf-butt
        hinges (9). Remove leaf-butt hinge.
    (b) Slide clamp (10) out of remaining leaf-butt hinge.
    (c) Place new clamp (10) into remaining leaf-butt hinge and slide removed leaf-butt hinge (9) on clamp.
    (d) Position leaf-butt hinge on rack and secure with two screws (8), flat washers (7), and nuts (6).

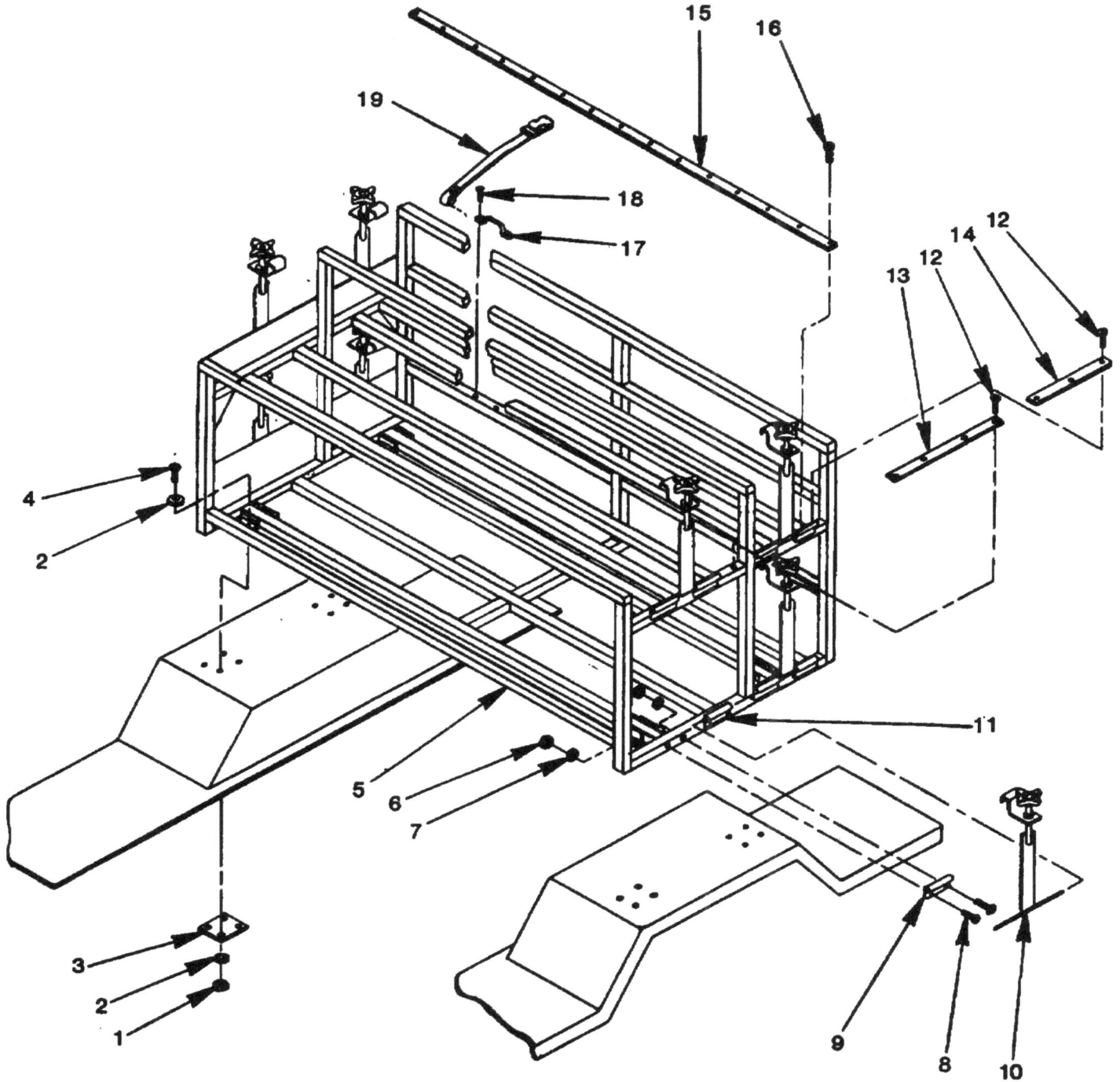

Figure 4-13. Stowage Rack Assembly Repair and Replacement on AN/MJQ-32.

(2) Runners.

    (a) Remove eight screws (12) from long clamp runner (13); or remove six screws (12) from short clamp runner (14).

    (b) Remove 15 screws (16) from runner (15) and take off runner.

    (c) Position new runner (15) on rack assembly (5) and secure with 15 screws (16).

    (d) Install long clamp runners (13) and secure with eight screws (12) or install short clamp runners (14) and secure with six screws (12).

(3) Strap Fastener.

    (a) Drill out solid rivets (18) securing strap fastener (17) to rack assembly (5).

    (b) Remove strap webbing (19) from strap fastener.

    (c) Place new strap fastener (17) through strap webbing (19) loop and position strap fastener on rack assembly and secure with solid rivets (18).

    (d) Touch up with paint as required.

    b. Removal. Power Plant AN/MJQ-32 contains a stowage rack for transporting the antenna accessories. (See figure 4-13.)

(1) Remove 16 locknuts (1, figure 4-13), 32 washers (2), 4 plates (3), and 16 screws (4).

(2) Remove rack (5) from trailer.

    c. Installation.

(1) Position rack (5) on trailer.

(2) Install 16 screws (4), 16 flat washers (2), 4 plates (3), 16 flat washers (2), and 16 locknuts (1).

(3) Tighten hardware and secure rack to trailer.

Section XVI. BRACKETS/HOLDERS/SUPPORTS AND GROUND STUD MAINTENANCE

4-27. FUEL CAN BRACKET REPLACEMENT.

This task covers: 1. Removal

                                      2. Installation

Initial Setup:
1. Tools - General Mechanics Tool Kit (5180-00-177-7033)
2. Materials/Parts - Fuel Can Bracket (96906); MS53052-1

WARNING

Before performing any maintenance that requires climbing on or under trailer, set trailer handbrakes, chock both wheels, and lower rear leg prop. Injury to personnel could result from trailer suddenly rolling or tipping.

There are two fuel can brackets supplied with the AN/MJQ-32 and four fuel can brackets supplied with the AN/MJQ-33. One bracket on the AN/MJQ-32 is mounted on each fender extension; one, roadside rear, near the switch box; the other, curbside front, facing front of generator. Two brackets on the AN/MJQ-33 are mounted on each side of the front generator; the other two are mounted in front of accessory box. Replacement procedures described below are typical for both power plant brackets. The AN/MJQ-33 power plant is shown.

  a. Removal.

(1) Remove four screws (1, figure 4-14), nuts (2), and flat washers (3) securing bracket (4) to fender extension (AN/MJQ-32) or bed (AN/MJQ-33).

(2) Remove fuel can bracket (4) from fender extension or bed.

  b. Installation.

(1) Position fuel can bracket (4) on fender extension (AN/MJQ-32) or bed (AN/MJQ-33).

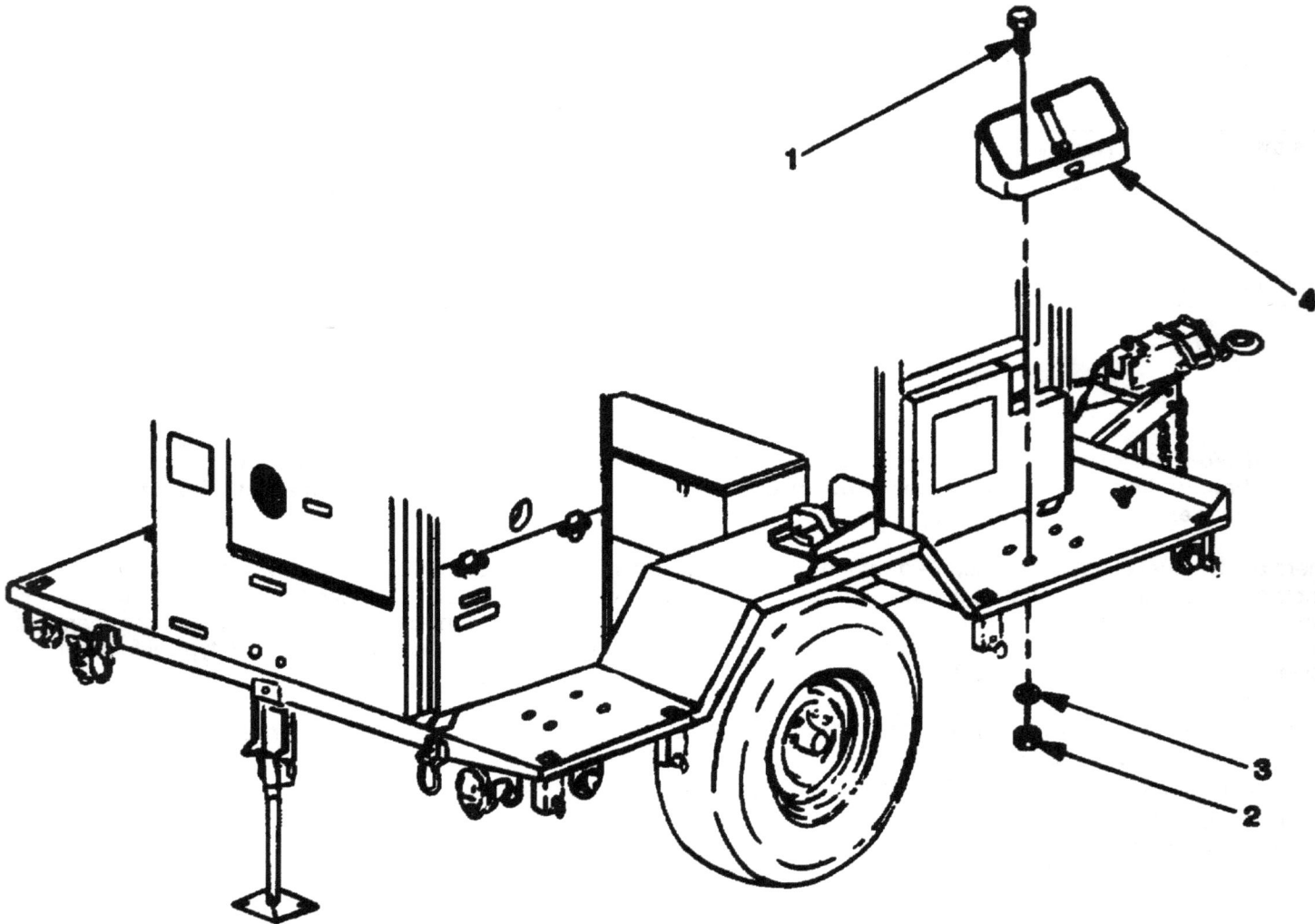

Figure 4-14. Fuel Can Bracket Replacement.

(2) Insert four screws (1) down through bracket (4) and through fender extension or bed.

(3) Install one washer (3) and nut (2) on each screw (1). Tighten hardware to secure bracket.

4-28. FIRE EXTINGUISHER BRACKET REPLACEMENT.

This task covers: 1. Removal

2. Installation

Initial Setup:

1. Tools - General Mechanics Tool kit (5180-00-177-7033)

2. Materials/Parts - Fire Extinguisher Bracket (97403); 13214E1235

Each fire extinguisher supplied with the power plants is carried in a bracket.
The bracket is mounted on the fender on the roadside of AN/MJQ-32 and curbside for AN/MJQ-33. The bracket is mounted upright on the AN/MJQ-32 and on its side for the AN/MJQ-33. Replacement procedures described below are the same for each bracket of either power plant. Shown is the bracket for AN/MJQ-33.

   a. Removal.

(1) Remove four screws (1, figure 4-15), flat washers (2) and nuts (3) securing bracket (4) to fender (5).

(2) Remove bracket (4) from fender (5).

   b. Installation.

(1) Position fire extinguisher bracket (4) on fender (5).

(2) Insert four screws (1) through bracket and fender.

(3) Install one flat washer (2) and nut (3) on each screw (1). Tighten hardware to secure bracket.

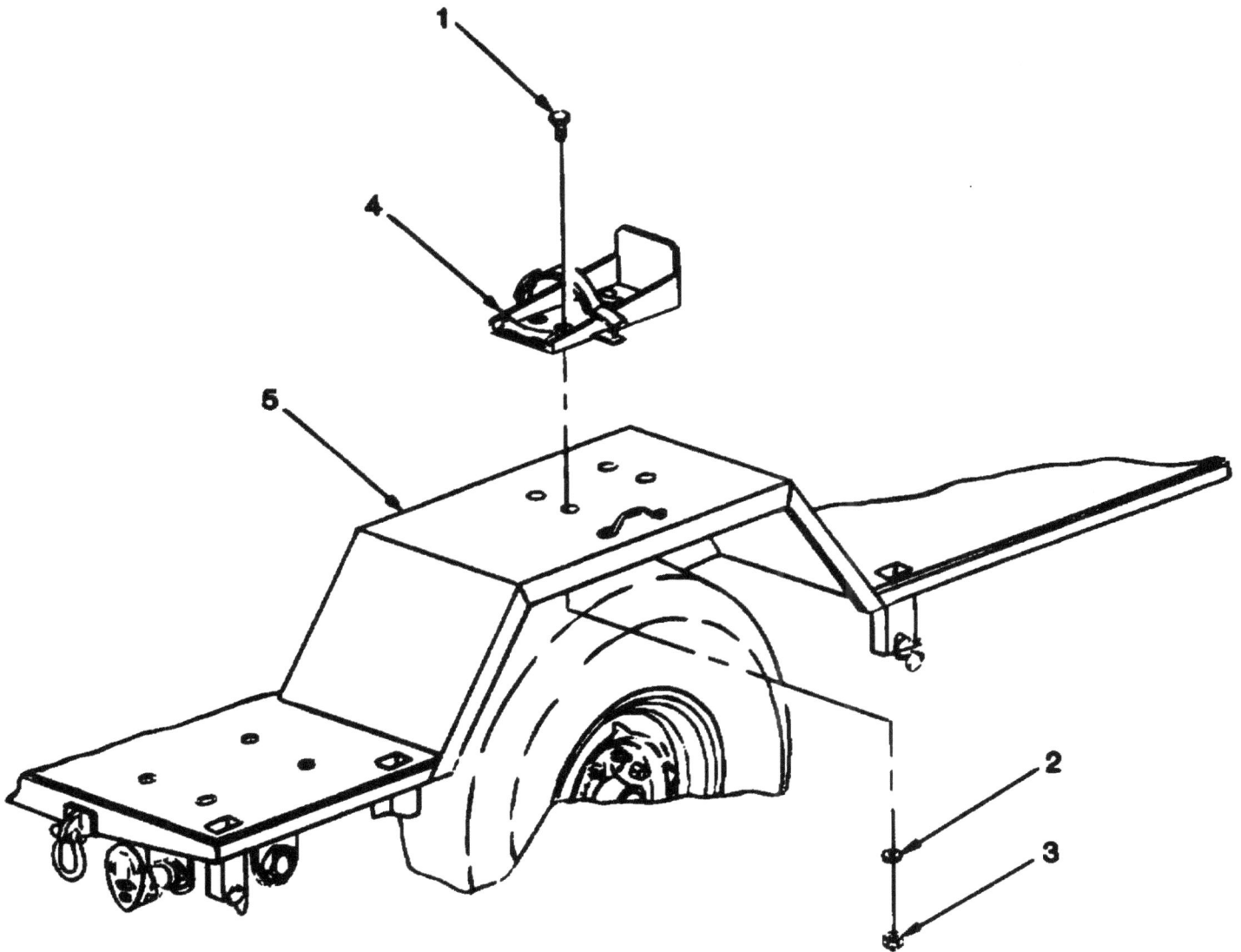

Figure 4-15. Fire Extinguisher Bracket Replacement.

4-29. ANTENNA MAST MOUNT REPLACEMENT, AN/MJQ-32 ONLY.

This task covers: 1. Removal

2. Installation

Initial Setup:
1. Tools - General Mechanics Tool Kit (5180-00-177-7033)
2. Materials/Parts - Mount (97403); 13228E9897-1

Each of the two antenna mast mounts is located on each side of the front generator.

a. Removal.

(1) Remove four locknuts (1, figure 4-16), eight flat washers (2), and four screws (3), securing mast mount (4) to front fender extension.

(2) Remove mast mount from front fender extension. b. Installation.

(1) Position mast mount (4) on front fender extension (5).

(2) Insert four screws (3) and four flat washers (2) through mast mount and front fender extension.

(3) Install four flat washers (2) and locknuts (1) on each screw (3). Tighten hardware to secure mast mount to fender extension.

4-30. ANTENNA MAST SUPPORT REPLACEMENT, AN/MJQ-32 ONLY.

This task covers: 1. Removal

2. Installation

Initial Setup:
1. Tools - General Mechanics Tool Kit (5180-00-177-7033)
2. Materials/Parts - Mount Support (97403); 13228E9897-2

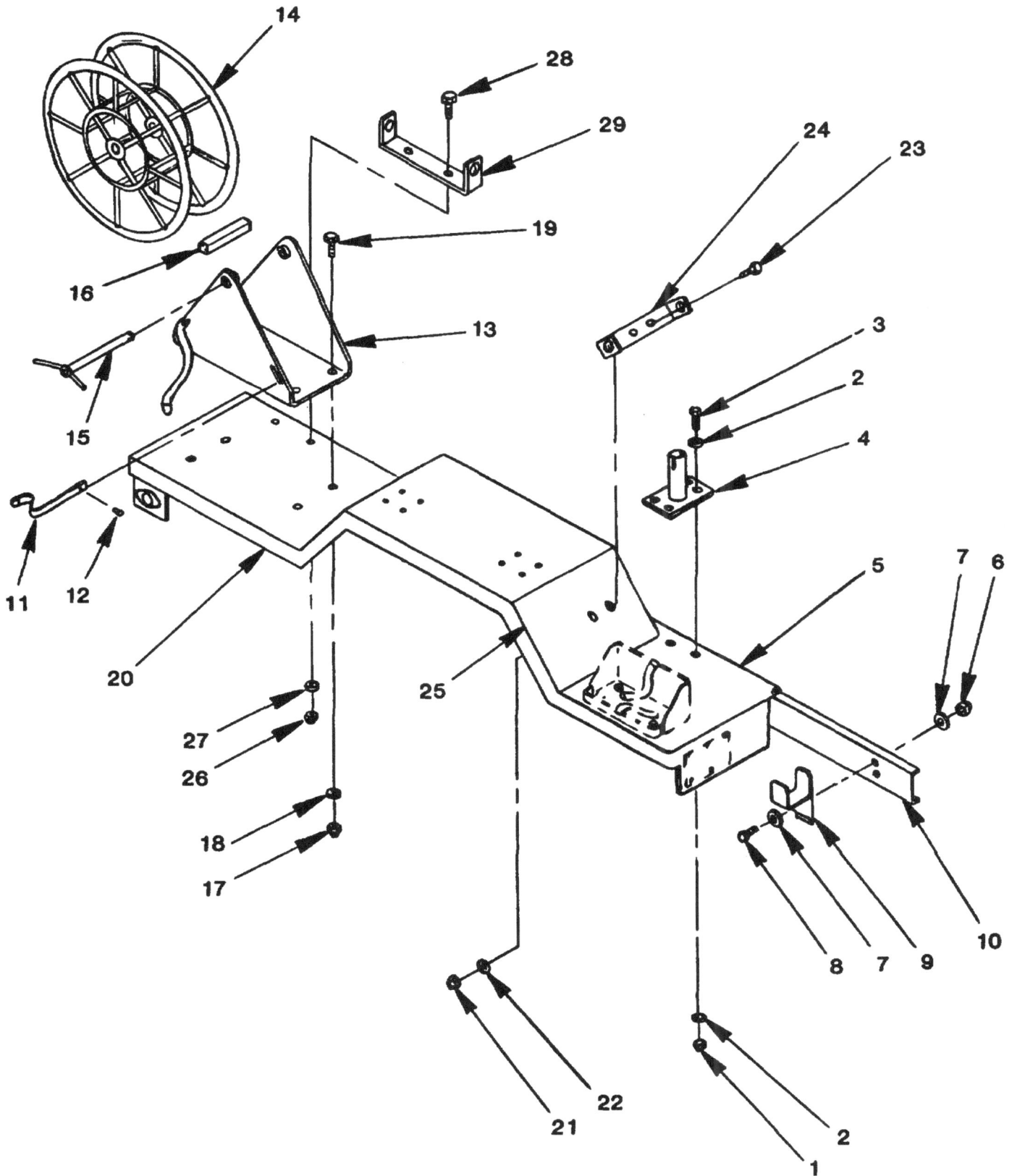

Figure 4-16. AN/MJQ-32, Curbside, Exploded View.

Each of the two antenna mast supports is located directly in front of the antenna mast mounts on the trailer chassis.

a. Removal.

(1) Remove two locknuts (6, figure 4-16), four flat washers (7), and two screws (8), securing mast support (9) to trailer frame (10).

(2) Remove mast support from trailer frame.

b. Installation.

(1) Position mast support (9) on trailer frame (10).

(2) Insert two screws (8) and two flat washers (7) through support and trailer frame.

(3) Install two flat washers (7) and locknuts (6). Tighten hardware to secure mast support.

4-31. CABLE-REEL BRACKET REPAIR AND REPLACEMENT, AN/MJQ-32 ONLY

This task covers: 1. Repair 3. Installation

2. Removal

Initial Setup:

1. Tools - General Mechanics Tool Kit (5180-00-177-7033)
   - Riveter (5120-00 -148-5847)
1. Materials/Parts - Cable-Reel Bracket (97403); 13217E2062
   - Rivets (96906); MS9319-208
   - Strapping (83149); MIL-W-530

The load-cable reel supplied with power plant AN/MJQ-32 is mounted in a cable-reel bracket. The bracket is located on the rear curbside fender extension.

a. Repair. Repair of the cable-reel bracket is by replacement of the strapping.

(1) Remove strapping (11, figure 4-16) by removing rivet (12) holding strapping to cable-reel bracket (13).

(2) Route one end of new strapping (11) through cable-reel bracket (13) slot and loop the strap end.

(3) Connect looped end with rivet (12).

b. Removal.

(1) Loosen strapping (11, figure 4-16) and remove from cable reel (14).

(2) Remove cable-reel from bracket (13) by loosening hold-down assembly (15) and removing assembly. Remove sleeving (16) from inside of cable reel.

(3) Remove four locknuts (17), flat washers (18), and screws (19) securing bracket (13) to rear fender extension (20).

(4) Remove cable-reel bracket from rear fender extension.

c. Installation.

(1) Position cable-reel bracket (13) on rear fender extension (20).

(2) Insert four screws (19) down through the bracket and through rear fender extension.

(3) Install one washer (18) and locknut (17) on each screw (19). Tighten hardware to secure bracket.

(4) Place sleeving (16) inside cable reel (14).

(5) Install cable reel on bracket (13) and slide hold-down assembly (15) through reel and bracket and tighten.

(6) Secure cable reel (14) by tightening strapping (11).

4-32. SPOUT, CAN, HOLDER REPLACEMENT, AN/MJQ-32 ONLY.

This task covers: 1. Removal

2. Installation

Initial Setup:
  1. Tools - General Mechanics Tool Kit (5180-00-177-7033)
  2. Materials/Parts - Spout Holder (97403); 13212E3553-2

The spout, can, holder is mounted on the curbside front section of the fender.

    a. Removal.

    (1) Remove two locknuts (21, figure 4-16), flat washers (22), and screws (23) securing spout bracket (24) to front section of fender (25).

    (2) Remove bracket from fender.

    b. Installation.

    (1) Position bracket (24) on front section of fenaer (25).

    (2) Insert two screws (23) through bracket and fender.

    (3) Install one flat washer (22) and locknut (21) on each screw.

    (4) Tighten hardware to secure bracket.

4-33. FUEL ADAPTER HOLDER REPLACEMENT, AN/MJQ-32 ONLY.

This task covers: 1. Removal

                    2. Installation

Initial Setup:

    1. Tools - General Mechanics Tool Kit (5180-00-177-7033)

    2. Materials/Parts - Fuel-Adapter Holder (97403); 13212E3553-1

The fuel-adapter holder is located, curbside, on the rear fender extension alongside the cable-reel bracket.

    a. Removal.

    (1) Remove two locknuts (26, figure 4-16), flat washers (27), and screws (28) securing holder (29) to curbside section of rear fender extension (20).

    (2) Remove holder from fender extension.

b. Installation.

(1) Position holder (29) on curbside section of rear fender extension (20).

(2) Insert two screws (28) through holder and rear fender extension.

(3) Install one flat washer (27), locknut (26) on each screw. Tighten hardware to secure holder.

4-34. HAMMER BRACKET ASSEMBLY REPLACEMENT, AN/MJQ-32 ONLY.

This task covers: 1. Removal

2. Installation

Initial Setup:

    1. Tools - General Mechanics Tool Kit (5180-00-177-7033)

    2. Materials/Parts - Hammer Bracket (97403); 13212E3553-1

The hammer bracket assembly is located at the rear of the trailer.

  a. Removal.

(1) Loosen the two wingnuts (1, figure 4-17) which hold bracket plate (2) in place and allow bracket plate to swing downward.

(2) Remove locknut (3), two flat washers (4), and screw (5) securing bracket assembly (6) to rear chassis (7).

(3) Remove bracket assembly from rear chassis.

  b. Installation.

(1) Position bracket assembly (6) on rear chassis (7).

(2) Insert screw (5) and flat washer (4) through bracket assembly and chassis.

(3) Install flat washer (4) and locknut (3) on bolt. Tighten hardware to secure bracket assembly.

(4) Rotate bracket assembly plate (2) to upright position and tighten wingnuts (1).

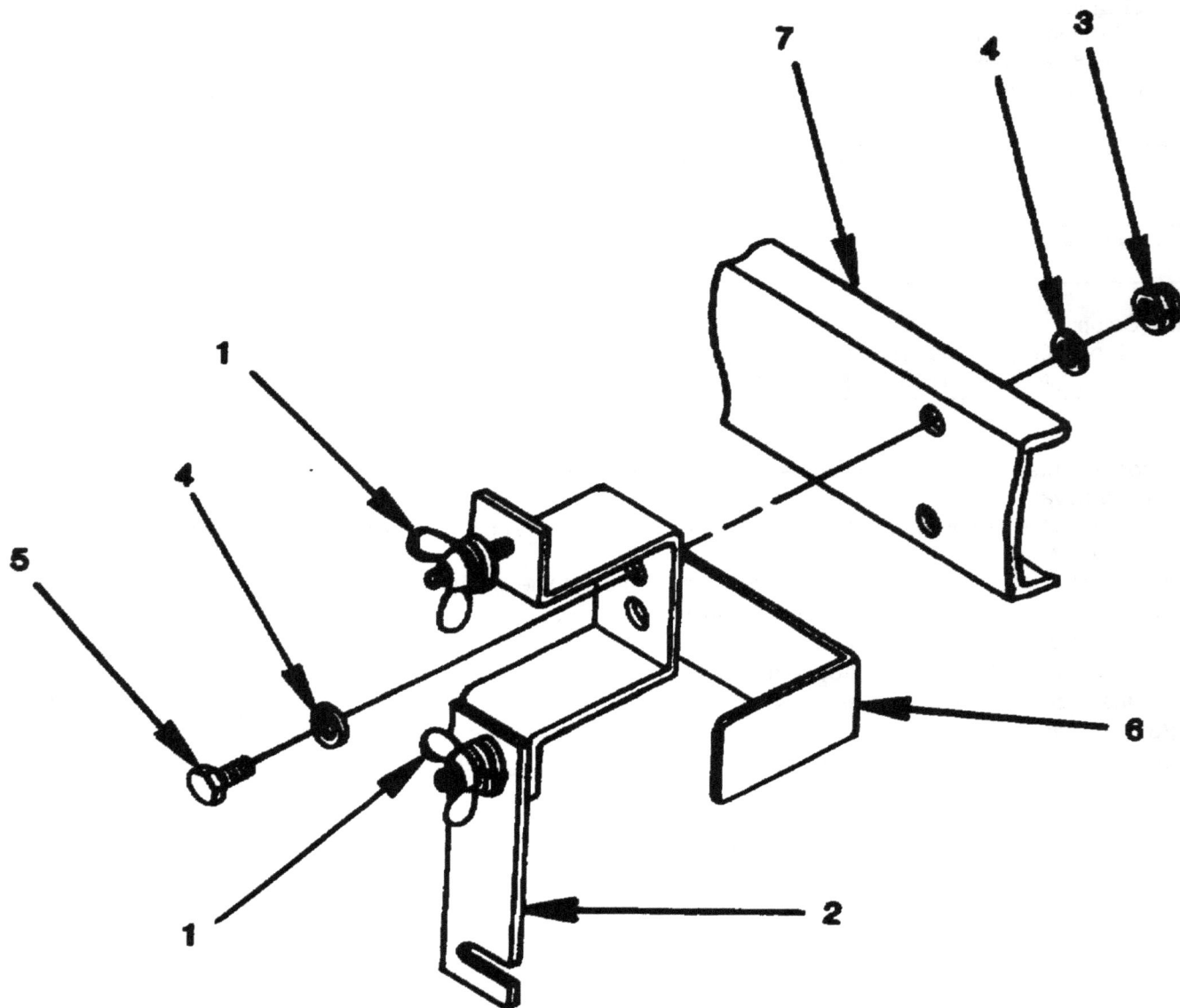

Figure 4-17. Hammer Bracket Assembly Replacement on AN/MJQ-32 Only.

4-35. IMPACT ROD/FUEL DRUM ADAPTER BRACKET ASSEMBLY REPLACEMENT, AN/MJQ-32 ONLY

This task covers: 1. Removal	
	2. Installation

Initial Setup:
    1. Tools - General Mechanics Tool Kit (5180-00-177-7033)
    2. Materials/Parts - Impact Rod Bracket (97403); 13212E3617

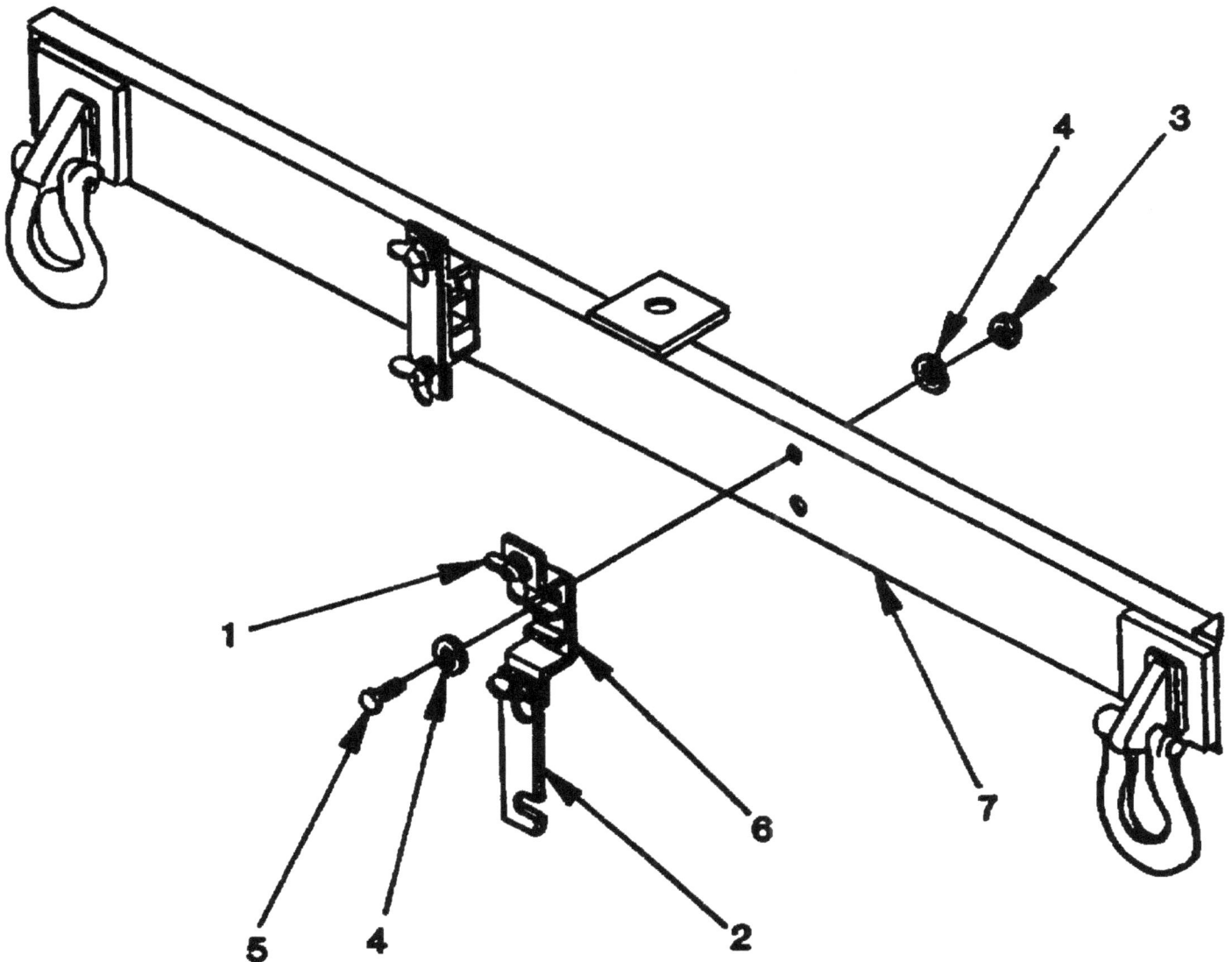

Figure 4-18. Impact Rod/Fuel Drum Adapter Bracket Assembly Replacement, AN/MJQ-32 Only.

The impact rod/fuel drum adapter bracket assembly is located in front of the trailer between the tow bars.

  a. Removal.

    (1) Loosen two wing nuts (1, figure 4-18) which hold bracket assembly plate (2) in place and allow plate to swing downward.

    (2) Remove two locknuts (3), four flat washers (4), and two screws (5) securing bracket assembly (6) to front chassis (7).

    (3) Remove bracket assembly from front chassis.

b. Installation.

(1) Position bracket assembly (6) on front chassis (7).

(2) Insert two screws (5) and two flat washers (4) through bracket assembly (6) and front chassis (7).

(3) Install flat washer (4) and locknut (3) on each screw (5). Tighten hardware to secure bracket assembly (6).

(4) Rotate bracket assembly plate (2) to upright position and secure by tightening wingnuts (1).

4-36. DRIVER/PULLER HOLDER REPLACEMENT, AN/MJQ-32 ONLY.

This task covers: 1. Removal

2. Installation

Initial Setup:

    1. Tools - General Mechanics Tool Kit (5180-00-177-7033)

    2. Materials/Parts - Driver/Puller Holder (97403); 13228E9898

The driver/puller holder is located on roadside front fender extension.

a. Removal.

(1) Remove two locknuts (1, figure 4-19), four flat washers (2), and two screws (3) securing holder (4) to front fender extension (5).

(2) Remove holder from front fender extension.

b. Installation.

(1) Position holder (4) on front fender extension (5).

(2) Insert two screws (3) and two flat washers (2) through holder (4) and front fender extension (5).

(3) Install flat washer (2) and locknut (1) on each screw (3). Tighten hardware to secure holder.

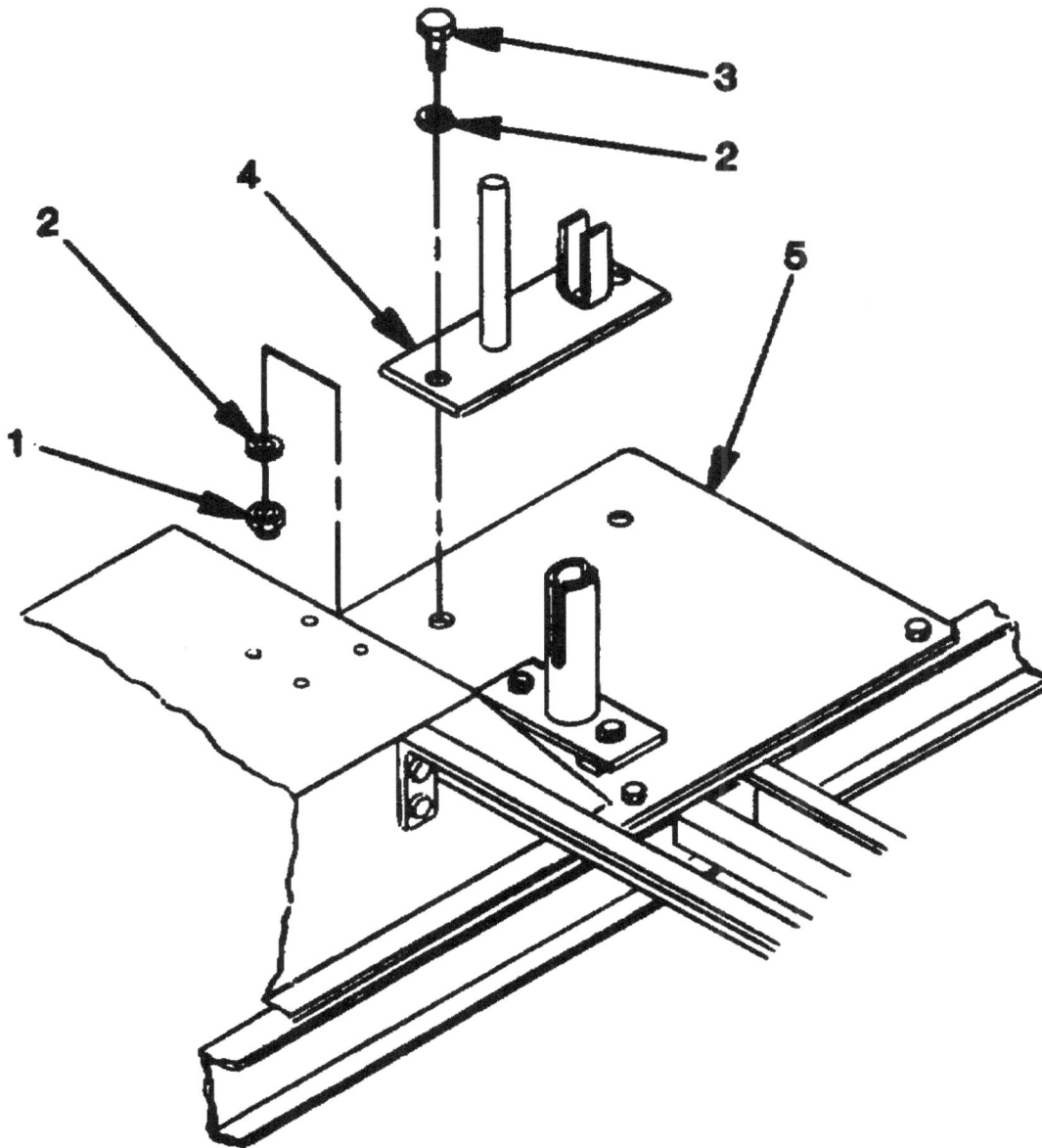

Figure 4-19. Driver/Puller Holder Replacement on AN/MJQ-32.

4-37. SWITCH BOX BRACKET REPLACEMENT.

This task covers: 1. Removal
2. Installation

Initial Setup:
1. Tools - General Mechanics Tool Kit (5180-00-177-7033)
2. Materials/Parts - Switch Box Bracket (97403); 13229E2303-2 (AN/MJQ-32)
13229E2303-1 (AN/MJQ-33)

The switch boxes, one on each power plant, are located on the roadside rear fender extension for the AN/IMQ-32 and on the roadside fender for the AN/MJQ-33.

Replacement procedures are the same for both power plants. Shown is the switch box for AN/MJQ-33.

a. Removal.

(1) Remove switch box (paragraph 4-19c(1)).

(2) Remove six locknuts (9, figure 4-20) (if AN/MJQ-33) or four locknuts (if AN/MJQ-32), flat washers (10), and screws (11) securing bracket (d) to roadside fender (if AN/MJQ-33), or roadside rear fender extension (if AN/MJQ-32).

(3) Remove bracket from fender or fender extension.

(4) Remove wingnut (1), washer (2), ground wire (3), washer (4), nut (5), and star washer (6) securing ground stud (7) to bracket (8).

b. Installation.

(1) Position bracket (8) on fender or fender extension.

(2) Insert six screws (if AN/MJQ-33) or four screws (if AN/MJQ-32) (11) through bracket (8) and roadside fender, or roadside rear fender extension.

(3) Install flat washer (10) and locknut (9) on each screw (11).
    Tighten hardware to secure bracket.

(4) Insert ground stud (7) into bracket (8) frame and install star washer (6), nut (5) (tighten), washer (4), ground wire (3), washer (2), and wingnut (1). Tighten wingnut.

(5) Install switch box (paragraph 4-19c(2)).

4-38. POWER CABLE BRACKET REPLACEMENT.

This task covers: 1. Removal

　　　　　　　　　　　　　　2. Installation

Initial Setup:

　　1. Tools - General Mechanics Tool Kit (5180-00-177-7033)

　　2. Materials/Parts - Power Cable Bracket (97403); 1312E3612

The power cable brackets are used to route the power cabling between the generator sets and the switch box for both power plants. They are situated at various locations along the cable path and are common to both power plants.

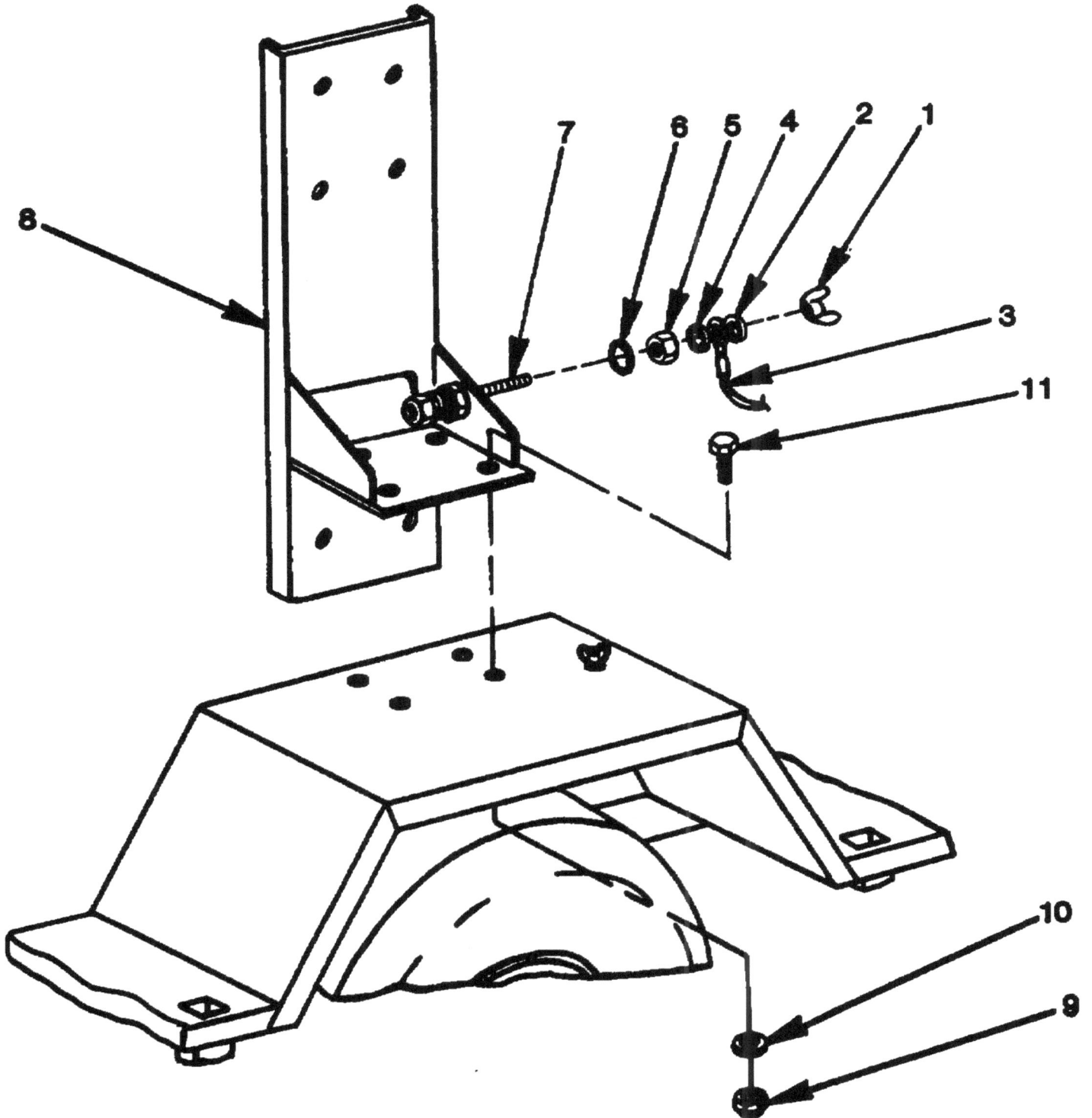

Figure 4-20. Switch Box Bracket Replacement.

a. Removal.

(1) Remove locknut (1, figure 4-21), flat washer (2), and screw (3) securing bracket (4) to trailer (5).

(2) Remove bracket from trailer.

b. Installation.

(1) Install bracket (4) on trailer (5).

(2) Install screw (3), washer (2) and locknut (1).

(3) Tighten hardware to secure bracket.

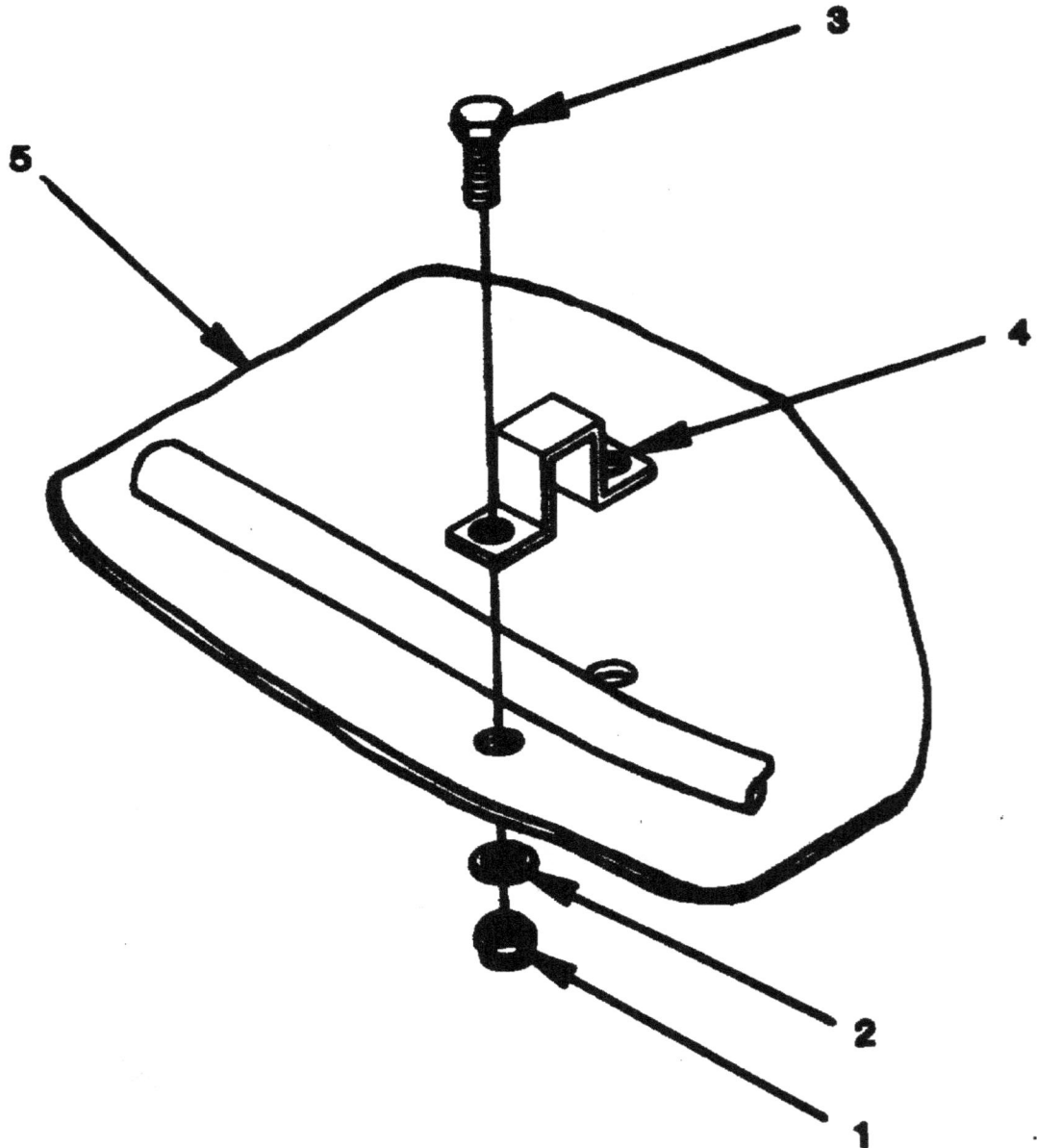

Figure 4-21. Switch Box Power Cable Bracket Replacement.

4-39. GROUND STUD REPLACEMENT.

This task covers: 1. Removal

2. Installation

Initial Setup:
1. Tools - General Mechanics Tool Kit (5180-00-177-7033)
2. Materials/Parts - Ground Stud (97403); 13214E1223

The ground studs, two on each power plant, are located in the same general area for either power plant. Stud A is located on the roadside rear area and stud B is located on the curbside front area. Both A studs are identical; also, both B studs are identical. Replacement procedures are the same for both power plants.

a. Stud A Removal. (See figure 4-22).

(1) Remove wingnut (1), two washers (2), large nut (3), washer (4), three ground terminals (5), washer (6), small nut (7), and star washer (8),

(2) Remove stud (9) from trailer and remove star washer (10) and nut (11).

b. Installation.

(1) Install small nut (11) and star washer (10) on stud (9) and insert into trailer.

(2) Install star washer (8) and small nut (7). Tighten nuts (7, 11).

(3) Install washer (6), three ground terminals (5), washer (4), and large nut (3). Tighten nut.

(4) Install two washers (2) and wingnut (1). Tighten wingnut.

c. Stud B Removal. (See figure 4-22).

(1) Remove wingnut (1), two washers (2), nut (3), washer (4), two ground terminals (5), washer (6), nut (7), and star washer (8).

(2) Remove stud (9) from trailer and remove star washer (10) and nut (11).

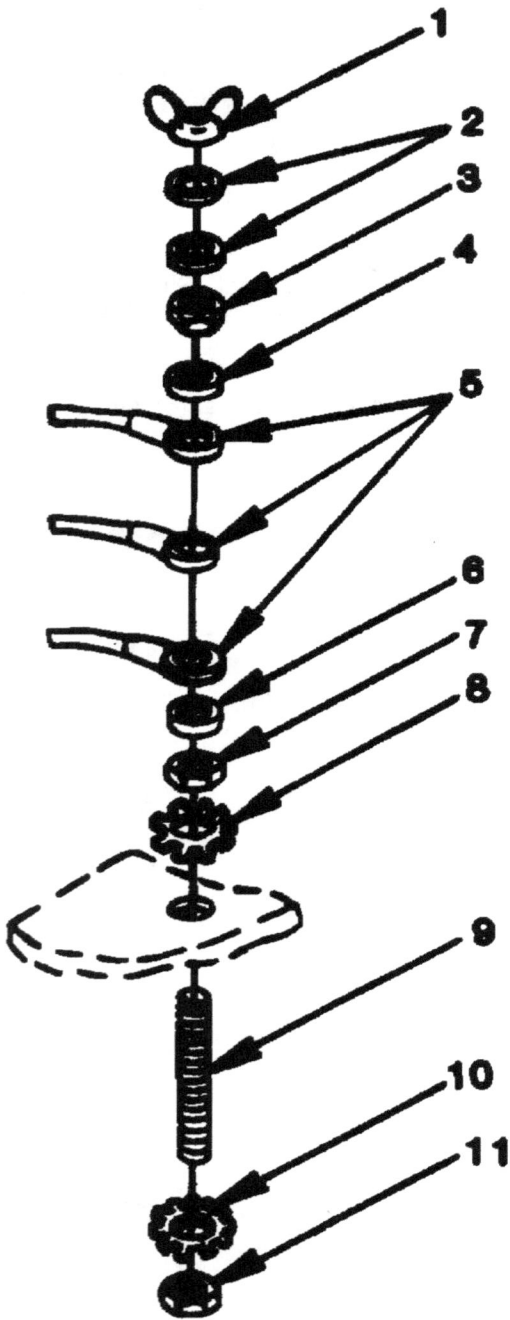

GROUND STUD A

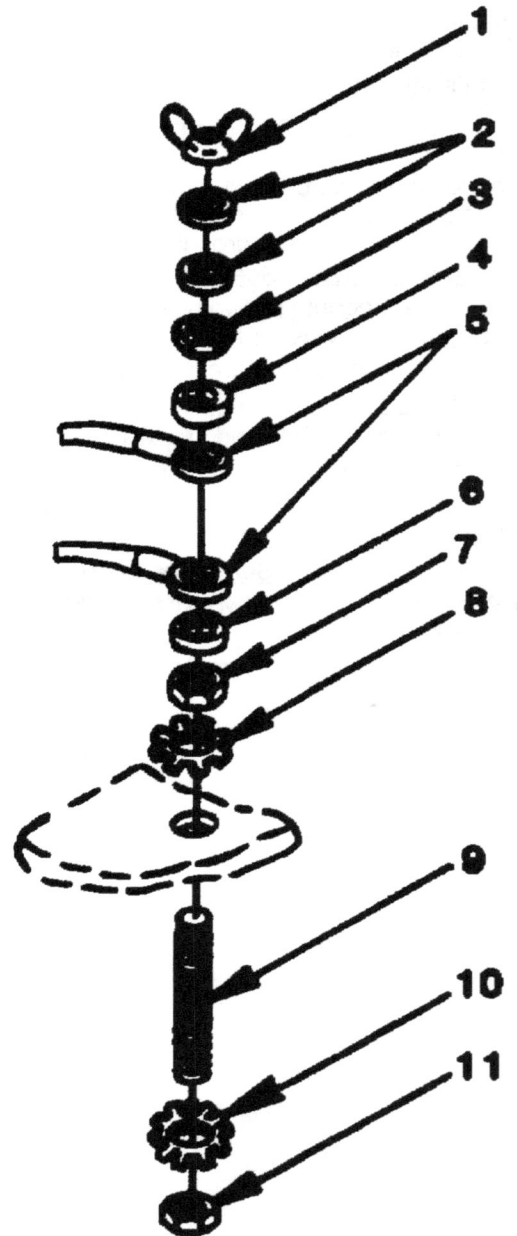

GROUND STUD B

Figure 4-22. Ground Stud Replacement.

d. Installation.
(1) Install nut (11) and star washer (10) on stud (9) and insert into trailer.
(2) Install star washer (8) and nut (7). Tighten nuts (7, 11).
(3) Install washer (6), two ground terminals (5), washer (4), and nut (3). Tighten nut.
(4) Install two washers (2) and wingnut (1). Tighten wingnut (1).

Section XVII. TRAILER ASSEMBLY MAINTENANCE
4-40. GENERAL.
The trailer assembly, for unit maintenance purposes, consists of leg prop and the taillight cable assembly and electrical lead.

WARNING

Before performing any maintenance that requires climbing on or under trailer, set trailer handbrakes, chock both wheels, and lower rear leg prop. Injury to personnel could result from trailer suddenly rolling or tipping.

4-41. LEG PROP MAINTENANCE.

This task covers: Servicing

Initial Setup:
    1. Tools - General Mechanics Tool Kit (5180-00-177-7033)
    2. Materials/Parts - Solvent, PD-680 (81348); (6850-00-664-5685)
        - GAA Grease (81349); (9150-00-190-0904)

For information on servicing the rear leg prop assembly see paragraph 4-9b.

4-42. TAILLIGHT CABLE ASSEMBLY AND ELECTRICAL LEAD MAINTENANCE.

This task covers: 1. Test 3. Repair

2. Removal 4. Installation

Initial Setup:
 1. Tools - General Mechanics Tool Kit (5180-00-177-7033)
  - Multimeter AN/PSM-45 (6625-01-139-2512)
 1. Materials/Parts - Connectors, Male (96906); MS27142-2
  - Connectors, Female (96906); MS27142-1
  - Tape, Electrical

The taillight cable assembly and electrical leads have been modified (lengthened) to compensate for the repositioning of the right and left rear stop lights. Therefore, for the tasks covering testing, removal, repair, and installation, refer to TM9-2330-202-14&P.

Section XVIII. PREPARATION FOR SHIPMENT AND STORAGE

4-43. PREPARATION FOR SHIPMENT.

The power plants AN/MJQ-32 and AN/MJQ-33 can be shipped by rail, air, or sea without damage to the units. The power plants will be packaged for shipment in accordance with MIL-G-28554B, Level A, B, or Commercial packing instructions.

4-44. PREPARATION FOR STORAGE

 a. Placement of equipment in administrative storage should be for short periods of time when a storage of maintenance effort exists. Items should be in mission readiness within 24 hours or within the time factors as determined by the directing authority. During the storage period, appropriate maintenance records will be kept.

 b. Before placing equipment in administrative storage, current maintenance services and equipment serviceable criteria (ESC) evaluations should be completed; shortcomings and deficiencies should be corrected, and all modification word orders (MWO's) should be applied.

 c. Storage site selection. Inside storage is preferred for items selected for administrative storage. If inside storage is not available, trucks, vans, conex containers and other containers may be used.

CHAPTER 5
DIRECT SUPPORT AND GENERAL SUPPORT MAINTENANCE INSTRUCTIONS

Section I. REPAIR PARTS, SPECIAL TOOLS, TEST, MEASUREMENT, AND DIAGNOSTIC EQUIPMENT (TMDE), AND SUPPORT EQUIPMENT

5-1. GENERAL.

This chapter contains Direct Support and General Support level maintenance procedures for components of the AN/MJQ-32 and AN/MJQ-33 which are not part of the basic generator sets or trailers. For all other maintenance procedures on the trailer, refer to TM9-2330-202-14&P. For maintenance procedures on the generator set, refer to TM5-6115-615-34.

5-2. SPECIAL TOOLS AND EQUIPMENT.

Test, Measurement ana Diagnostic Equipment (TMDE) and support equipment are listed in Appendix B, Section II. Refer to TM5-6115-615-24P for generator set TMDE and TM9-2330-202-14&P for trailer TMDE.

5-3. REPAIR PARTS.

Repair parts are listed and illustrated in the Repair Parts and Special Tools List (RPSTL) Appendix C.

5-4. FABRICATED TOOLS AND EQUIPMENT.

No fabricated tools and equipment are required for direct support maintenance on the power plants AN/MJQ-32 and AN/MJQ-33.

Section II. MAINTENANCE OF GENERATOR SETS

5-5. GENERATOR SET REPLACEMENT.

| This task covers: | 1. Removal |
| | 2. Installation |

Initial Setup:
    1. Tools - General Mechanics Tool Kit (5180-00-177-7033)
    2. Materials/Parts - Generator Set (30554); MEP701A

Power Plants AN/MJQ-32 and AN/MJQ-33 each have two MEP-701A generator sets.
Replacement procedures for either generator set are identical for both power plants, except where noted.

   a. Removal.

(1) Remove tarpaulin from AN/MJQ-33 (paragraph 3-6a(1)).

(2) Remove four bow assemblies from AN/MJQ-33 (paragraph 3-6b(1)).

(3) Remove ground wires from generator set ground stud (1, figure 5-1).

(4) Open service access door (2), and disconnect all load cables (3) and remove from generator set (4). Close access door.

(5) Remove four screws (5), eight washers (6), and four locknuts (7) securing generator set (4) to trailer bed (8).

(6) Remove engine oil drain hose (9) from grommet in AN/MJQ-33 trailer bed.

(7) Disconnect auxiliary fuel line (10), if connected.

WARNING

When lifting generator set, use lifting equipment with a minimum lifting capacity of 1500 lb. Do not stand under generator while it is being lifted. Failure to observe these precautions can case death or injury to personnel or damage to equipment.

(8) Open lifting eye access door (1, figure 5-2) and attach lifting equipment with a minimum lifting capacity of 1500 lb (2) to lifting eye (3) on top of generator set (4) and remove generator set from trailer.

   b. Installation.

WARNING

When lifting generator set, use lifting equipment with a minimum lifting capacity of 1500 lb. Do not stand under generator while it is being lifted. Failure to observe these precautions can cause death or injury to personnel or damage to equipment.

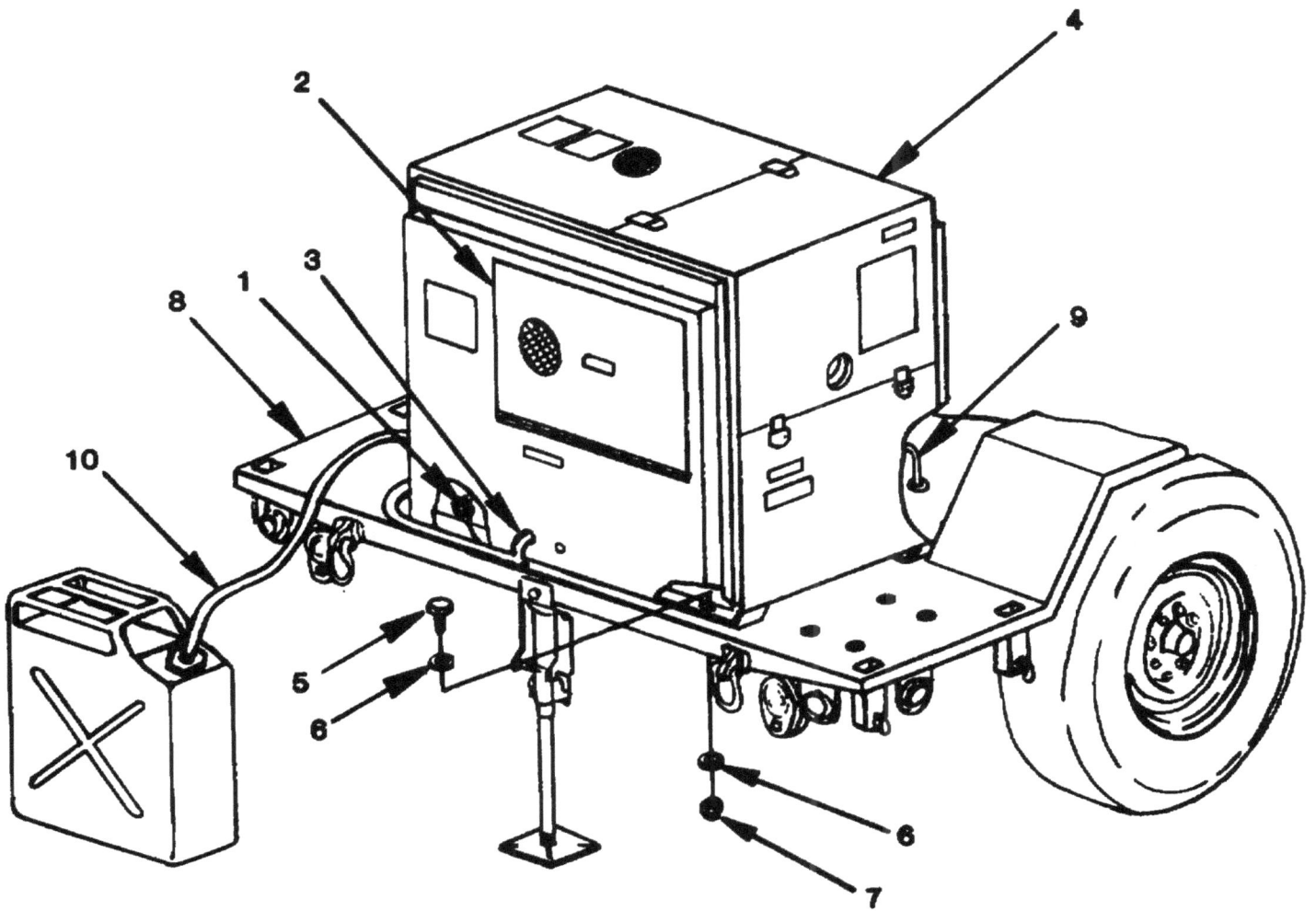

Figure 5-1. Detaching Generator Set from Trailer.

(1) Open lifting eye access door (1, figure 5-2) and attach lifting equipment with a minimum lifting capacity of 1500 pounds (2) to lifting eye (3) on top of generator set (4) and lift generator.

(2) Lower generator set (4, figure 5-1) on trailer bed (8) and align mounting holes.

(3) Insert four screws (5), four washers (6) through generator set and trailer.

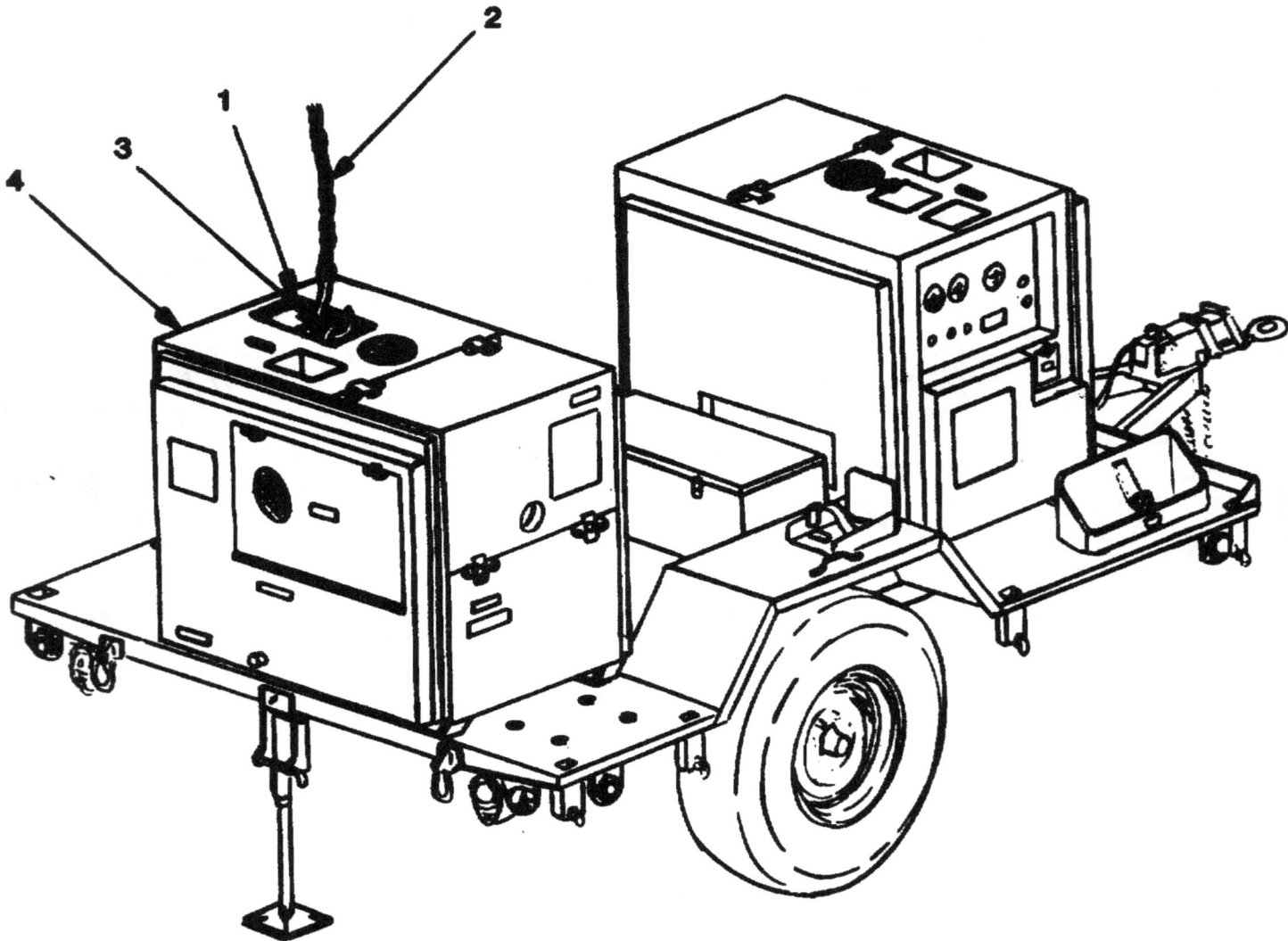

Figure 5-2. Lifting Generator Set from Trailers.

(4) Install four washers (6) and tighten locknuts (7).

(5) Insert engine oil drain hose (9) through grommet in AN/MJQ-33 trailer bed (8).

(6) Install four bow assemblies, AN/MJQ-33 (paragraph 3-6b(1)).

(7) Install tarpaulin, AN/MJQ-33 (paragraph 3-6a(1)).

Section III. MAINTENANCE OF SWITCH BOX

5-6. <u>SWITCH REPLACEMENT.</u>

This task covers: 1. Removal

2. Installation

Initial Setup:
1. Tools - General Mechanics Tool Kit (5180-00-177-7033)
2. Materials/Parts - Switch (97403); 13219E9860

WARNING
Make sure generator sets are shut down before performing any maintenance on switch. Failure to observe this precaution may result in injury or death by electrocution.

NOTE
Compare replacement switch with original switch. Note locations and markings on terminals.

a. <u>Removal.</u>

(1) Remove 16 screws (1, figure 5-3), lockwashers (2) and flat washers (3) securing cover (4) to switch box (5) and pull cover away from switch box.

NOTE
Make sure identification bands on wires are legible before disconnecting wires from switch. Tag switch terminals and all unmarked wires.

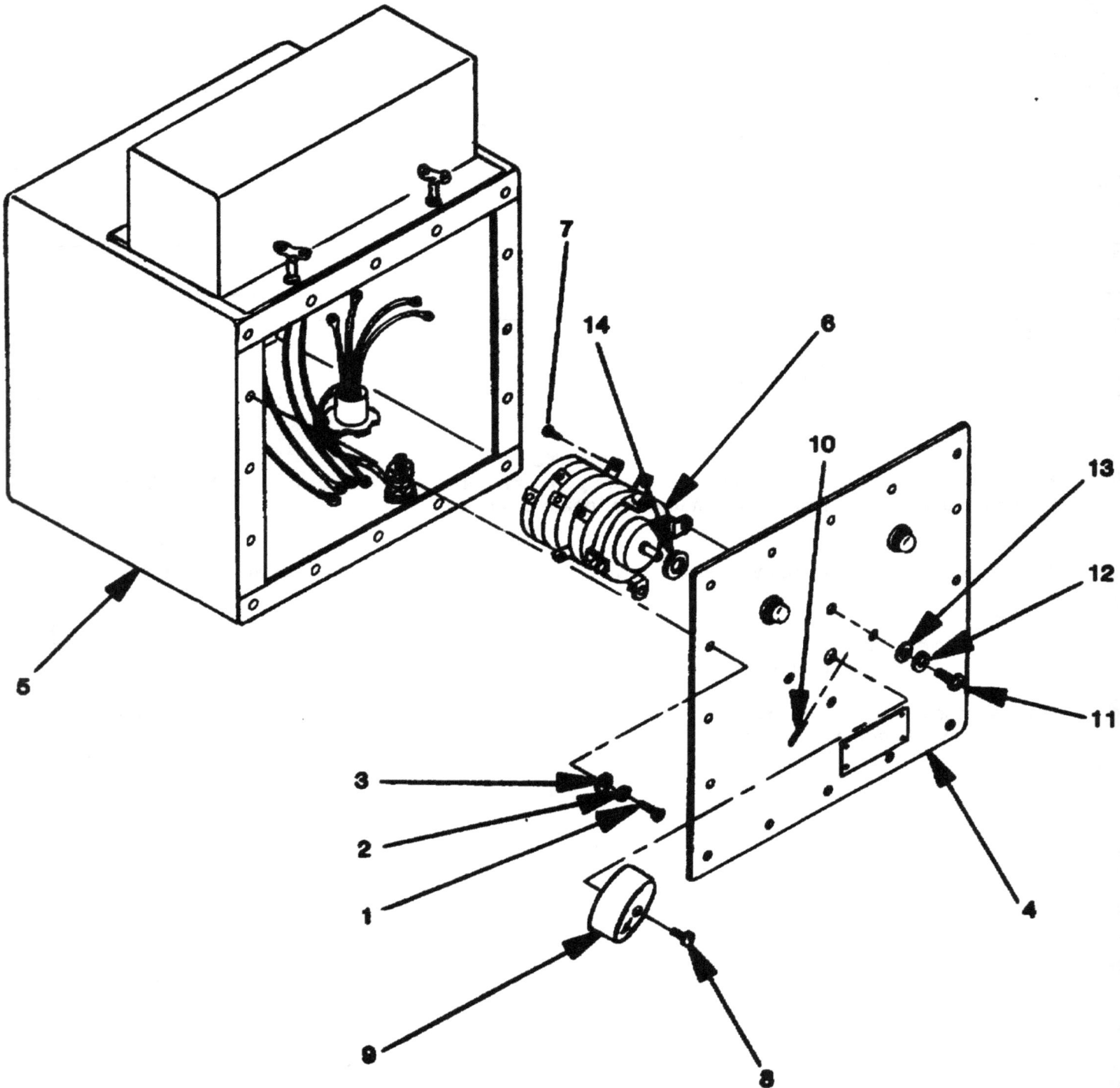

Figure 5-3. Switch Box Switch Replacement.

(2) Disconnect all wires from switch (6) by removing screws (7) from each switch terminal.

(3) Remove screw (8) securing knob (9) to switch (6) and take off knob.

(4) Unscrew pin (10) from switch shaft.

(5) Remove four screws (11), flat washers (12), and sealing washers (13) securing switch (6) to cover (4) and remove switch from cover.

(6) Remove sealing washer (14) from switch.

NOTE
Ensure that terminals B1 and B3 are facing upward when installing switch on cover.

b. Installation.

(1) Install sealing washer (14) on switch (6) and position switch on inside of cover (4). Install four sealing washers (13), flat washers (12), and screws (11) securing switch to cover (4).

(2) Screw pin (10) into switch shaft.

(3) Position knob (9) on switch (6) and install screw (8).

NOTE
Observe identification bands on wires when reconnecting wires.

(4) Connect indicator light wires to switch (paragraph 4-23d).

(5) Connect power cables to switch (paragraph 4-24b(2)).

(6) Connect load terminal wires to switch (paragraph 4-22b(2)).

(7) Position cover (4) on switch box (5) and secure with 16 flatwashers (3) lockwashers (2) and screws (1).

5-7. WIRING AND POWER CABLES MAINTENANCE .

WARNING
Make sure generator sets are shut down before performing any maintenance on wires or cables. Failure to follow this precaution may result in injury or death by electrocution.

Repair of the wiring and power cables of the switch box is by replacing loose or damaged terminals and wrapping exposed wires with electrical tape (item 7, Appendix F). Replacement terminals are soldered on the wires in accordance with procedures given in TB SIG 222 and TM 55-1500-323-25.

Section IV. MAINTENANCE OF ACCESSORY BOX, AN/MJQ-33

5-8. ACCESSORY BOX REPAIR, AN/MJQ-33 ONLY.

This task covers: Repair

Initial Setup:

1. Tools - General Mechanics Tool Kit (5180-00-177-7033)
   - Drill, 1/4-inch (5130-00-807-3009)
   - Riveter, Blind Head (5120-00-148-5847)
2. Materials/Parts - Catches (96906); MS18015-1
   - Strap Fasteners (96906); MS51939-3
   - Hinged Hasp/Staple (96906); MS27969-4
   - Rivets (96906); MS9460-IU2

The accessory box is repaired by replacing the hasp, the catches, and the strap fasteners, as required. The box itself may be straightened, welded, and repainted. If required, repaint in accordance with MIL-T-704 and MIL-C-46168.

a. *Catch and Hasp Replacement.*

(1) Remove accessory box (paragraph 4-21a(1)) from the trailer bed, if necessary.

(2) Drill out rivets (1, figure 5-4) securing hinged hasp/staple (2) or clamping/strike catch (3) to accessory box (4).

(3) Position new hinged hasp/staple (2) or clamping/strike catch (3) on accessory box (4) and secure with rivets (1).

(4) Touch up with paint as required.

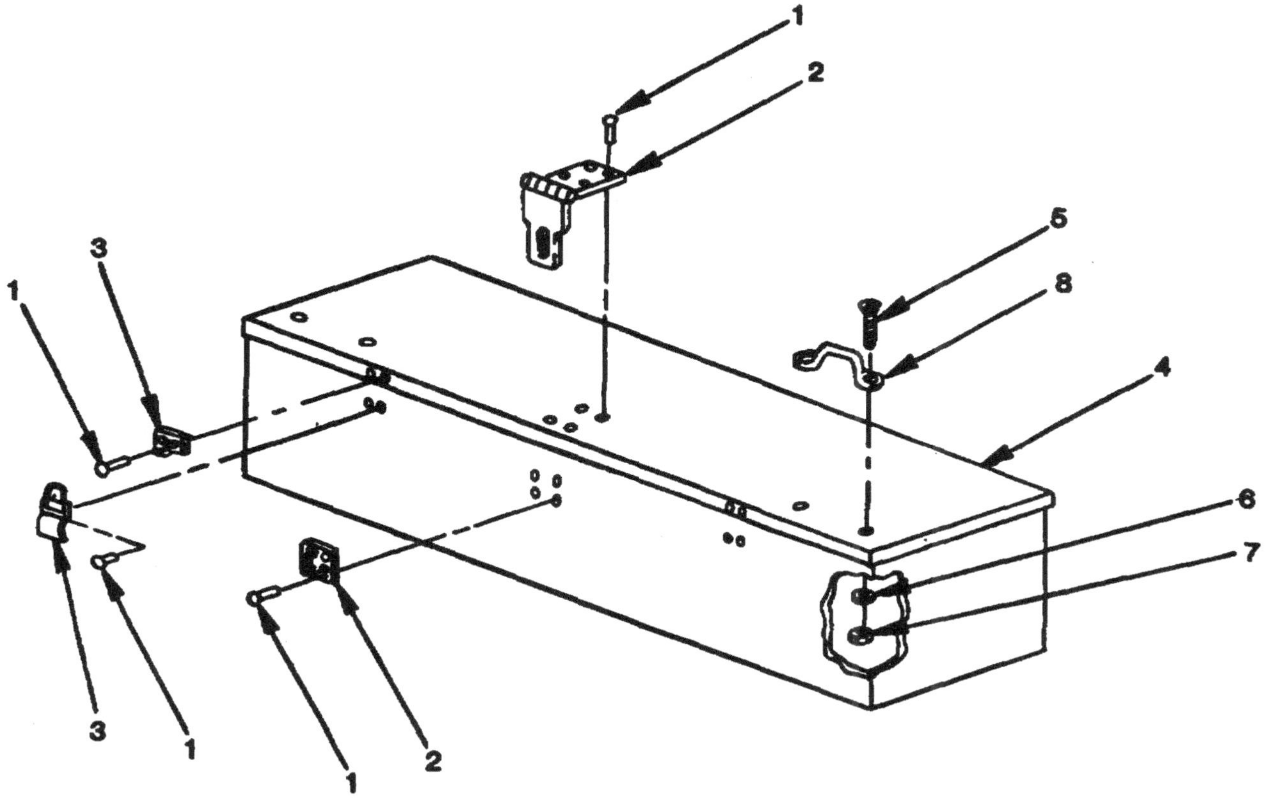

Figure 5-4. Accessory Box Repair on AN/MJQ-33.

b. Strap Fastener Replacement.

    (1) Remove two screws (5, figure 5-4), flat washers (6), and lock-nuts (7) securing strap fastener (8) to accessory box (4).

    (2) Position strap fastener (8) on accessory box (4) and install two screws (5), washers (6), and locknuts (7).

Section V. MAINTENANCE OF POWER PLANT TRAILERS

5-9. TRAILER ASSEMBLY REPLACEMENT .

This task covers 1. Removal

                           2. Installation

Initial Setup:

    1. Tools - General Mechanics Tool Kit (5180-00-177-7033)

    2. Materials/Parts - Trailer Assembly (974U3); 13228E9896 (AN/MJQ-32)

                                           13229E2302 (AN/MJQ-33)

WARNING
When lifting generator set, use lifting equipment with a minimum lifting capacity of 1500 lb. Do not stand under generator sets while they are being lifted. Failure to observe these precautions can cause injury or death to personnel or damage to equipment.

The removal and installation procedures for the generator sets are the same for either power plant. The stowage rack associated with the AN/MJQ-32 may have already have been removed by unit maintenance. If not, refer to paragraph 4-26 b(1).

    a. Removal.

        (1) Perform generator set Removal procedures given in paragraph 5-5a.

        (2) Remove stowage rack, AN/MJQ-32, if necessary (paragraph 4-26b(1)).

        (3) Remove trailer assembly.

b. Installation.

(1) Perform generator set Installation procedures given in paragraph 5-5b for installment on replacement trailer.

(2) Install stowage rack (paragraph 4-26b(2)) if previously removed in a(2) above.

NOTE

For follow-on-maintenance, refer to unit maintenance, Chapter 4, and general instructions (paragraph 5-1).

5-10. FENDER/FENDER EXTENSION REPLACEMENTS, AN/MJQ-32.

This task covers: 1. Removal 3. Installation
        2. Repair

Initial Setup:

 1. Tools - General Mechanics Tool Kit (5180-00-177-7033)
 2. Materials/Parts - Fender/Fender Extension (97403); 13228E9304
          - Fender/Fender Extension, Curbside (97403); 13228E9901

WARNING

A minimum of two persons is required to move the fender/ fender extensions due to the bulk and weight of these items. Failure to do so could result in serious injury. When removing mounting hardware, use support or hold the fender/fender extensions so they will not drop.

 a. Removal. (See figure 5-5). The fender/fender extension of the modified trailer of the AN/MJQ-32 consists of a single weldment that includes both fender and fender extensions. The following procedures are the same for either the curbside or roadside fender/fender extension.

  (1) Removal (Curbside).

    (a) Remove stowage rack (paragraph 4-26b(1)).

    (b) Remove cable-reel bracket (paragraph 4-31b).

    (c) Remove fuel-adapter holder (paragraph 4-33a).

5-11

(d) Remove spout, can, holder (paragraph 4-31a).

(e) Remove antenna mast mount (paragraph 4-29a).

(f) Remove fuel can bracket (paragraph 4-27a).

(g) Remove power cable brackets (paragraph 4-38a).

(h) Remove 5 locknuts (1, figure 5-5), 10 flat washers (2), and 5 screws (3) securing inner edge of curbside fender/fender extension (15) to trailer frame (13).

(i) Remove four locknuts (4), eight flat washers (5), and four screws (6) securing cross braces (14) to curbside fender/ fender extension.

(j) Remove four locknuts (7), eight flat washers (8), and four screws (9) securing center portion of curbside fender/fender extension to trailer frame (13).

(k) Remove 6 locknuts (10), 12 flat washers (11), and 6 screws (12) securing remainder of curbside fender/fender extensions to trailer frame.

(l) Remove curbside fender/fender extension from trailer frame.

(2) Removal (Roadside).

(a) Remove stowage rack (paragraph 4-26b(1)).

(b) Remove antenna mast mount (paragraph 4-29a).

(c) Remove driver/puller holder (paragraph 4-36a).

(d) Remove fire extinguisher bracket (paragraph 4-28a).

(e) Remove switchbox (paragraph 4-19c(1)) and bracket (paragraph 4-37a).

(f) Remove power cable brackets (paragraph 4-38a).

(g) Remove fuel can bracket (paragraph 4-27a).

(h) Perform steps (h) thru (k) of Removal (Curbside), above, for roadside fender/fender extension (16).

(i) Remove roadside fender/fender extension from trailer frame.

b. Repair. Repair of the fender/fender extensions is limited to straightening, welding, or repainting. If required, repaint in accordance with MIL-T-704 and MIL-C-46168.

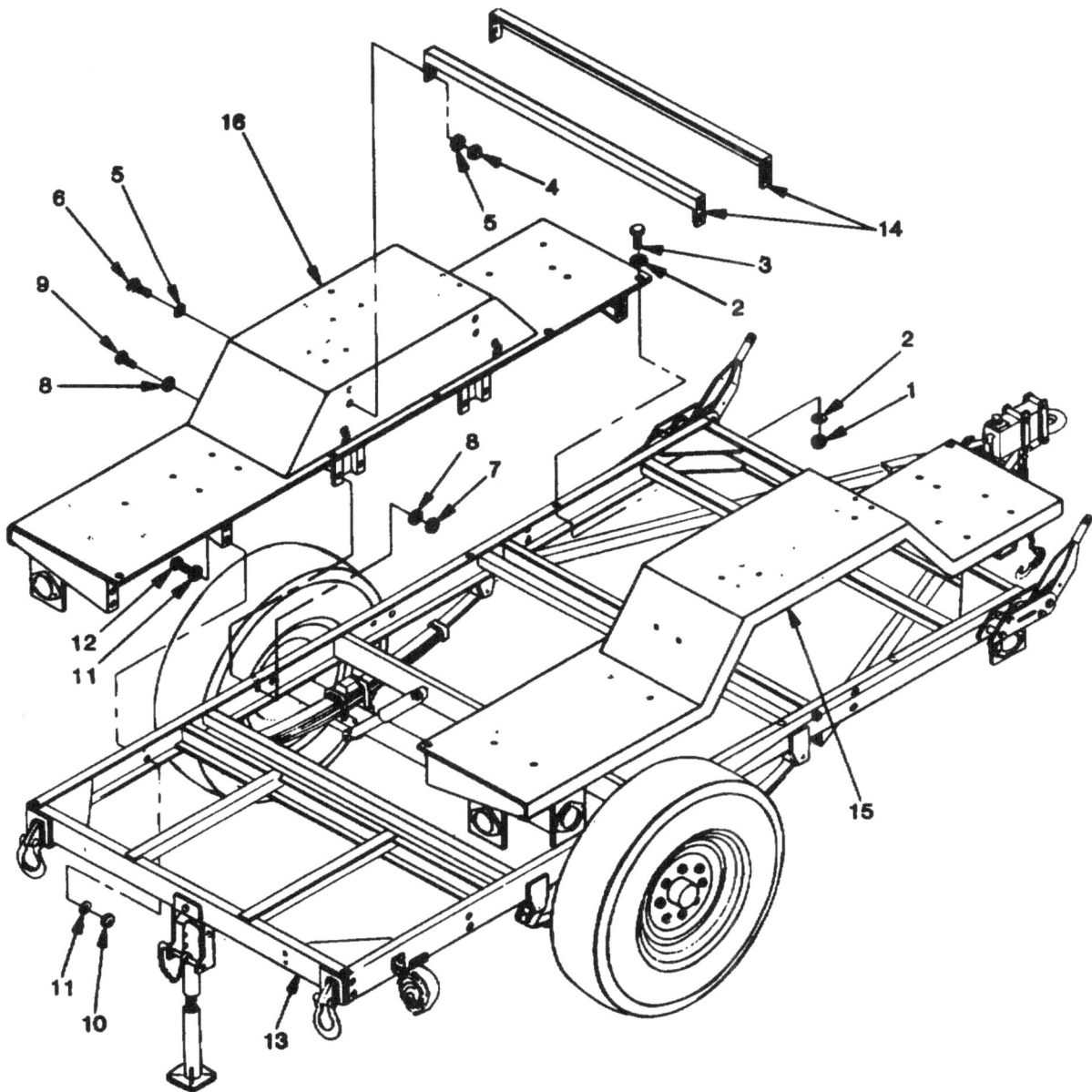

Figure 5-5. Fender/Fender Extension Replacement on AN/MJQ-32.

c. Installation.

(1) Installation (Curbside).

(a) Position fender/fender extension (15, figure 5-5) on trailer frame (13). Use support, if necessary.

(b) Install 5 screws (3), 10 flat washers (2), and 5 locknuts (1) and secure inner edge of fender/fender extension to trailer frame.

(c) Secure cross braces (14) to fender/fender extension with four screws (6), eight flat washers (5), and four locknuts (4).

(d) Install four screws (9) and four flat washers (8) through center portion of fender/fender extension and secure with four flat washers (8) and locknuts (7).

(e) Install six screws (12) and six flat washers (11) through remainder of fender/fender extension and secure with six flat washers (11) and locknuts (10).

(f) Install stowage rack assembly (paragraph 4-26b(2)).

(g) Install cable-reel bracket (paragraph 4-31c).

(h) Install fuel-adapter holder (paragraph 4-33b).

(I) Install spout, can, holder (paragraph 4-32b).

(j) Install antenna-mast mount (paragraph 4-30b).

(k) Install fuel can bracket (paragraph 4-27(2)).

(l) Install power cable brackets (paragraph 4-38b).

(2) Installation (Roadside).

(a) Perform steps a thru f in c(1) used in the above curbside installation, using fender/fender extension (16).

(b) Install antenna mast mount (paragraph 4-30b).

(c) Install driver/puller holder (paragraph 4-36b).

(d) Install fire extinguisher bracket (paragraph 4-28b).

(e) Install switchbox bracket (paragraph 4-37b) and switch box (paragraph 4-19c(2)).

(f) Install power cable brackets (paragraph 4-38b).

(g) Install fuel can bracket (paragraph 4-27b).

5-11. TRAILER BED AND FENDER REPAIR AND REPLACEMENT AN/MJQ-33.

This task covers: 1. Removal   3. Installation
                  2. Repair

Initial Setup:
1. Tools - General Mechanics Tool Kit (5180-00-177-7033)
2. Materials/Parts - Trailer Bed (97403); 13221E7326

    a. Removal. (See figure 5-6). The body of the modified AN/MJQ-33 trailer consists of a single weldment that includes both fender and the bed of the trailer.

(1) Remove tarpaulin (paragraph 3-6a(1)).

(2) Remove tarpaulin support and bow assembly (paragraph 3-6b(1)).

(3) Remove accessory box (paragraph 4-25a).

(4) Remove switchbox (paragraph 4-19c(1)).

(5) Remove generator sets (paragraph 5-5a).

(6) Remove 10 screws (1, figure 5-6), 20 flat washers (2), and 10 nuts (3) securing trailer body (4) to trailer chassis (5).

(7) Remove 16 screws (6), 32 flat washers (7), and 16 nuts (8) securing trailer body (4) to trailer chassis (5).

NOTE
Removal of the trailer body requires the removal and disassembly of both handbrake lever assemblies. The handbrake lever assemblies are symmetrical and this procedure is typical for both.

(8) Remove two screws (9), spacers (10), and nuts (11) and remove roadside handbrake lever assembly (12) from trailer chassis (5).

(9) Remove cotter pin (13), washer (14), shaft (15), and pulley (16) from handbrake lever assembly (12).

(10) Working under trailer, pull handbrake cable clevis (17) back through the holes in the front two braces on the trailer body.

(11) Repeat steps (8) thru (10) to remove and disassemble curbside handbrake lever assembly.

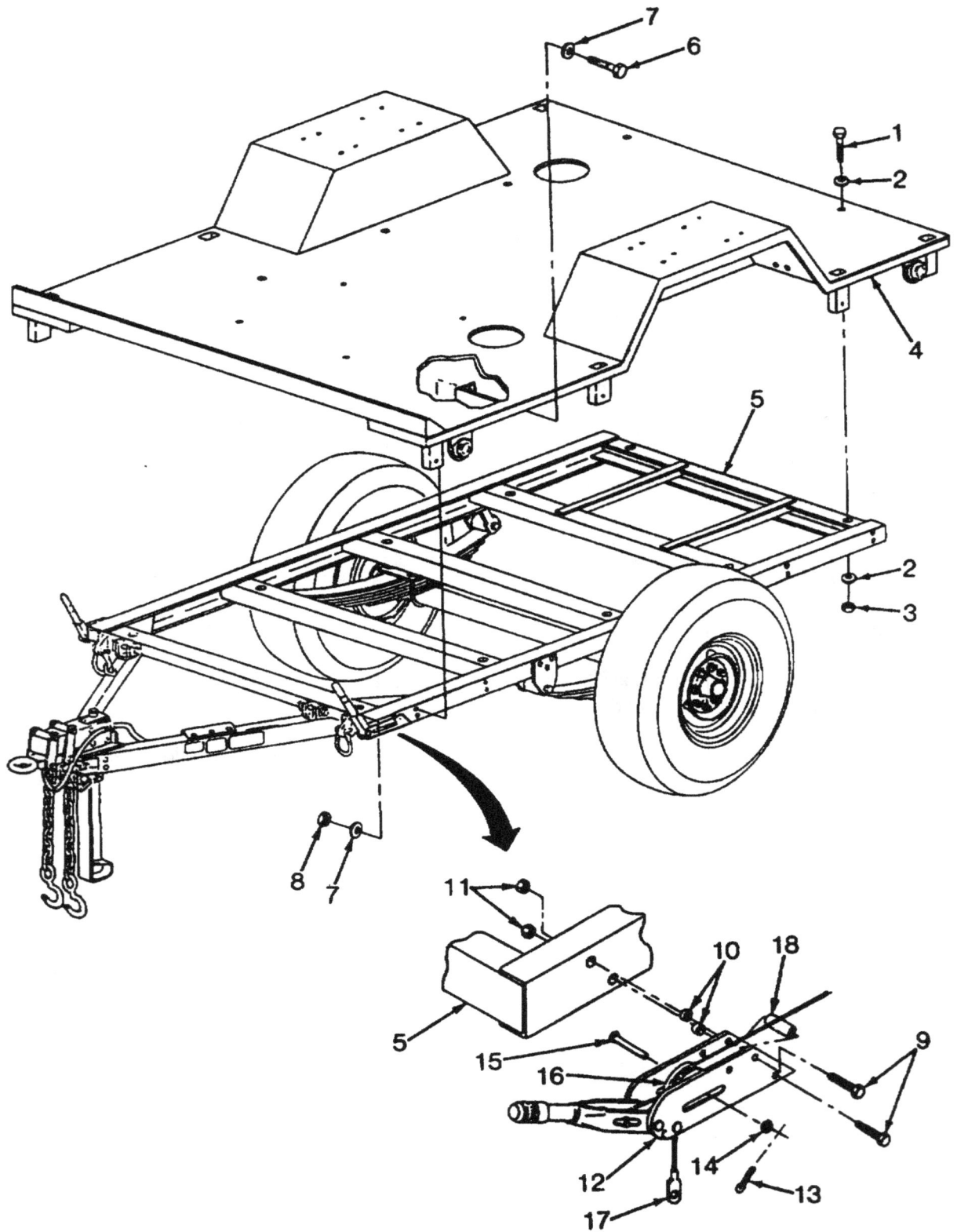

Figure 5-6. Fender and Bed Replacement on AN/MJQ-33.

WARNING

When lifting trailer body, use lifting equipment with a minimum capacity of 500 lb. Do not stand under trailer body while it is being lifted. Failure to observe these precautions can cause death or injury to personnel or damage to equipment.

(12) Using suitable lifting equipment with a minimum lifting capacity of 500 lb, lift trailer body (4) from trailer chassis (5).

b. Repair. Repair of the trailer bed and fenders is limited to straightening, welding, and repainting body. If required, repaint in accordance with MIL-T-704 and MIL-C-46168.

c. Installation.

WARNING

When lifting trailer body, use lifting equipment with a minimum capacity of 500 lb. Do not stand under trailer body while it is being lifted. Failure to observe these precautions can cause injury to personnel or damage to equipment.

(1) Using suitable lifting equipment with a minimum lifting capacity of 500 lb, left trailer body (4, figure 5-6) on trailer chassis (5) and align mounting holes.

(2) Insert 10 screws (1) with flat washers (2) through trailer bed and through trailer chassis frame rails.

(3) Working under trailer, install one flat washer (2) and nut (3) on each screw (1).

(4) Insert 16 screws (6) with flat washers (7) through trailer body braces into trailer chassis.

(5) Working under trailer, install one flat washer (7) and nut (8) on each screw (6). Tighten hardware to secure trailer body to trailer chassis.

(6) Feed roadside handbrake cable clevis (17) forward through holes in front two braces on trailer body.

(7) Wrap handbrake cable around pully (16), position pulley in handbrake lever assembly (12) and insert shaft (15).

(8) Install washer (14) and cotter pin (13) to secure shaft (15).

5-17

(9) Assemble handbrake lever assembly (12) using two screws (9) and spacers (10). Make certain top screw (9) goes through spacer (18) and bottom screw (9) goes through handbrake cable clevis.

(10) Position assembled handbrake lever assembly (12) on trailer chassis (5) and install two nuts (11). Tighten hardware.

(11) Repeat steps (6) thru (10) to assemble and install curbside handbrake lever assembly.

(12) Install generator set (paragraph 5-5b).

(13) Install accessory box (paragraph 4-25b).

(14) Install tarpaulin support and four bow assemblies (paragraph 3-6b(2).

(15) Install tarpaulin (paragraph 3-6a2).

5-12. REAR LEG PROP ASSEMBLY REPAIR AND REPLACEMENT.

This task covers: 1. Removal 3. Installation
                   2. Repair

Initial Setup:
  1. Tools - General Mechanics Tool Kit (5180-00-177-7033)
  2. Materials/Parts - Trailer Bed (97403); 13221E7326

Maintenance of the rear leg prop assembly consists of repairing or replacing the assembly as required. The rear leg prop assembly is the same for both power plants.

  a. Leg Prop Assembly Removal. (See figure 5-7.)

   (1) While supporting rear leg prop assembly, pull out angled bar (1, figure 5-7) and lower the leg prop assembly from its traveling position.

   (2) Line up boss (2) on upper leg (3) with holes in bracket (4) and insert angled bar (1) to lock leg in support position.

   (3) Remove either one of two cotter pins (5) from leg prop assembly pivot shaft (6).

   (4) While steadying leg prop assembly, remove shaft (6) with remaining cotter pin (5) in place.

5-18

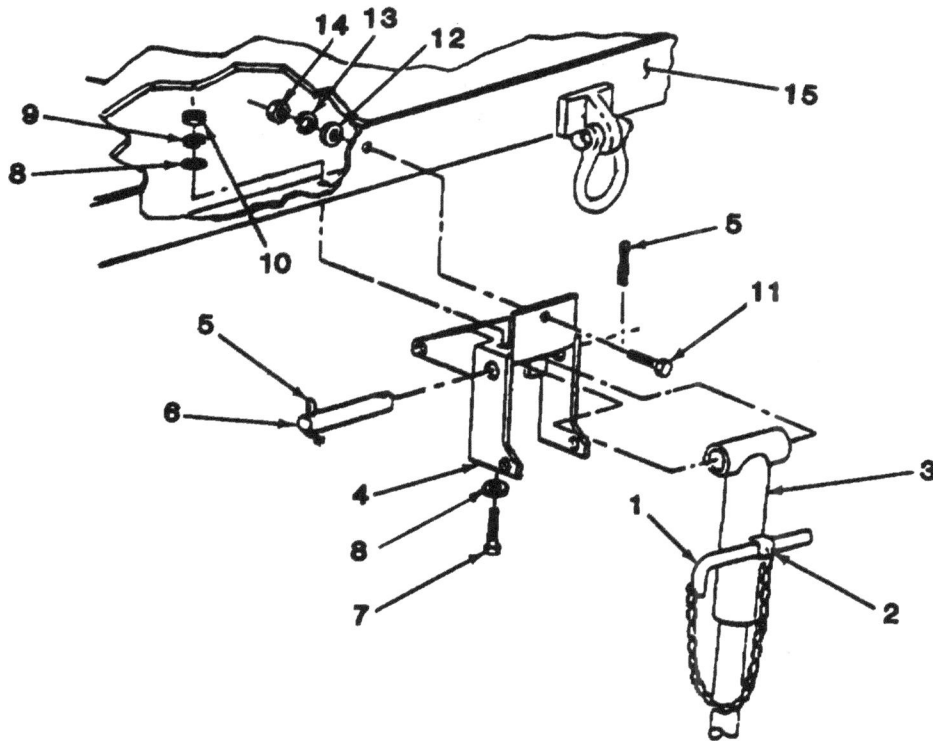

Figure 5-7. Rear Leg Prop Assembly Replacement.

WARNING

When angled bar is removed in step (5), rear leg prop assembly will fall from bracket if not support-ed. To prevent injury to personnel or damage to equipment, do not permit leg prop assembly to drop.

(5) Lift leg assembly slightly to take weight off angled bar (1) and remove bar.

(6) Lower leg assembly from bracket (4).

(7) Remove two screws (7), four flat washers (8), two lockwashers (9), and two nuts (10).

(8) Remove screw (11), flat washer (12), lockwasher (13) and nut (14). Remove bracket (4) from trailer frame (15).

b. Rear Leg Prop Assembly Repair. (See figure 5-8). Repair of the rear leg prop assembly is limited to welding and repainting. Partial disassembly may be necessary to facilitate repair. If required, repaint in accordance with MIL-T-704, and MIL-C-46168.

(1) Disassembly.

(a) Remove rear leg prop assembly from trailer (paragraph 5-12a(1)).

(b) Clamp rear leg prop assembly in vise with spring pin (1, figure 5-8) facing up.

(c) Remove drive spring pin from upper leg (2) and remove leg base (3).

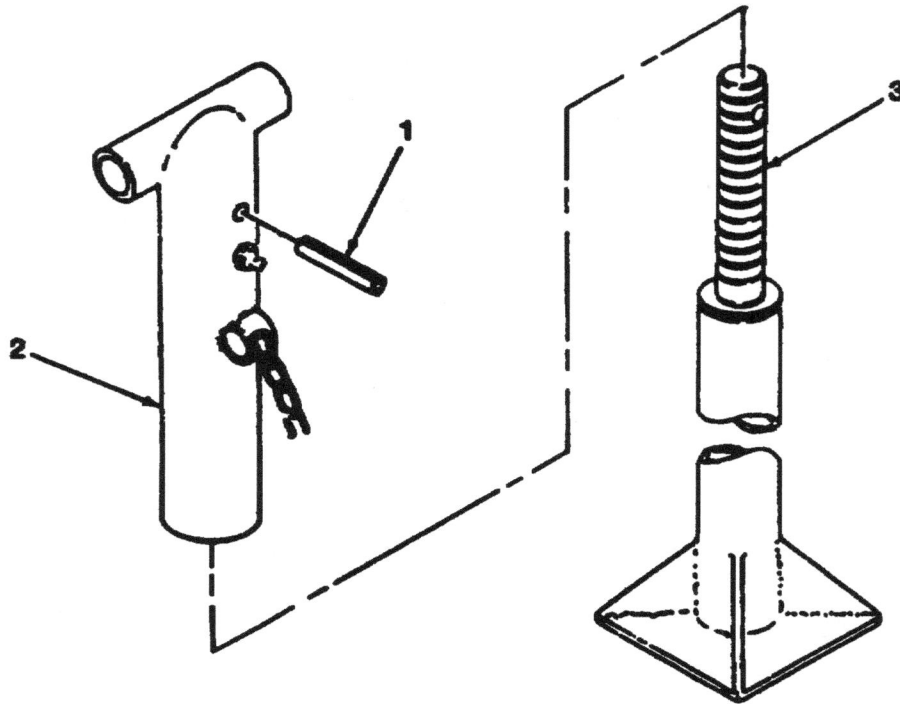

Figure 5-8. Rear Leg Prop Disassembly.

5-20

(2) Assembly.

    (a) Clamp upper leg (2) in vise with spring pin hole facing up.

    (b) Insert leg base (3) into upper leg and turn leg base until hole in screw lines up with hole in upper leg.

    (c) Install spring pin (1) to secure leg base to upper leg.

    (d) Install leg prop assembly on trailer (paragraph 5-12a(2)).

c. Rear Leg Prop Assembly Installation.

    (1) Position bracket (4, figure 5-7) on trailer frame (15) and install two screws (7), four flat washers (8), two lockwashers (9) and nuts (1U).

    (2) Install screw (11), flat washer (12), lockwasher (13), and nut (14). Tighten all hardware.

    (3) Lift rear leg prop assembly into bracket (4) and secure by inserting angled bar (1) through holes in bracket and boss (2) on upper leg (3).

    (4) Position rear leg prop assembly to line up boss on top of leg with pivot holes in bracket (4). Insert pivot shaft (6).

    (5) Insert cotter pin (5) in pivot shaft (6) and bend cotter pin legs in opposite directions.

    (6) Pull out angled bar (1) to unlock leg prop assembly.

    (7) Swing rear leg prop assembly up into traveling position and secure by inserting angled bar (1) through holes in bracket (4) and boss (2) on upper leg (3).

5-13. TARPAULIN REPAIR, AN/MJQ-33 ONLY .

Repairs to the tarpaulin shall be made in accordance with FM10-16, Fabric Repairing.

CHAPTER 6
TEST AND INSPECTION AFTER REPAIR

Section I. GENERAL REQUIREMENTS

6-1. GENERAL REQUIREMENTS.

The activity performing the repair is responsible for the performance of all applicable tests and inspections specified in the technical manuals referenced below. Activities performing maintenance on any component of the power plants must perform those tests and inspections required by the applicable component or system repair instruction.

Section II. INSPECTION

6-2. GENERATOR SET INSPECTIONS.

Refer to TM 5-6115-615-12 and -34 for inspections required following repair of the generator set.

6-3. TRAILER INSPECTIONS.

Refer to TM 9-2330-202-14&P for inspections required following. repair of the trailer.

Section III. OPERATIONAL TESTS

6-4. GENERATOR SET OPERATIONAL TESTS.

Refer to TM 5-6116-615-12 and -34 for operational tests required to verify satisfactory performance of the generator set.

6-5. TRAILER OPERATIONAL TESTS.

Refer to TM 9-2330-202-14&P for operational tests required to verify satisfactory performance of the trailer.

6-1/(6-2 blank)

APPENDIX A
REFERENCES

A-1. SCOPE.

This appendix lists all pamphlets, forms, technical manuals, specifications, and miscellaneous publications referenced in this manual.

A-2. FORMS AND RECORDS.

Recommended Changes to Publications and Blank Forms	DA Form 2028
Recommended Changes to Equipment Technical Publications	DA Form 2028-2
Depreservation Guide for Vehicles and Equipment	DA Form 2258
Equipment Inspection and Maintenance Worksheet	DA Form 2404
Maintenance Request	DA Form 5504
Consolidated Index of Army Publications	DA PAM 25-30
The Army Maintenance Management System (TAMMS)	DA PAM 738-750
Product Quality Deficiency Report	SF 368

A-3. MILITARY SPECIFICATIONS.

Chemical Agent Resistant Aliphatic Polyurethane Coating	MIL-C-46168
Identification Marking of U.S. Military Property	MIL-STD-130
Identification Marking of Combat and Tactical Transport	MIL-STD-642
Treatment and Painting of Materiel	MIL-T-704

A-4. TECHNICAL MANUALS.

NBC Decontamination	FM 3-5
Fabric Repairing, Tents, Canvas, Webbing	FM 10-16
Operator and Organizational Maintenance Manual for Generator Set, Diesel Engine Driven, Tactical Skid Mounted, 3 KW, 3 Phase, 120/208 Volts AC (DOD Model MEP-701A) Utility Class 60 HZ (NSN 6115-01-234-5966)	TM5-6115-615-12

Change 1 A-1

A-2

A-5. TECHNICAL BULLETINS.

A-3/(A-4 blank)

APPENDIX B
MAINTENANCE ALLOCATION CHART

Section I. INTRODUCTION

B-1. General

This appendix provides a summary of the maintenance operation for Power Plants AN/MJQ-32 and AN/MJQ-33. It authorizes categories of maintenance for specific maintenance functions on repairable items and components and the tools and equipment required to perform each function. This appendix may be used as an aid in planning Maintenance operations.

B-2. Maintenance Function

Maintenance shall be limited to the functions defined below:

a. Inspect. To determine the serviceability of an item by comparing its physical, mechanical, and/or electrical characteristics with established standards through examination.

b. Test. To verify serviceability and to detect incipient failure by measuring the mechanical or electrical characteristics of an item and comparing those characteristics with prescribed standards.

c. Service. Operations required periodically to keep an item in proper operating condition (i.e., to clean (decontaminate), to preserve, to drain, to paint, or to replenish fuel, lubricants, hydraulic fluids, or compressed air supplies.

d. Adjust. To maintain, within prescribed limits, by bringing into proper or exact position, or by setting the operating characteristics to the specified parameters.

e. Align. To adjust specified variable elements of an item to bring about optimum or desired performance.

f. Calibrate. To determine and cause corrections to be wade or to be adjusted on instruments or test measuring and diagnostic equipments used in precision measurement. Consists of comparisons of two instruments, one of which is a certified standard of known accuracy, to detect and adjust any discrepancy in the accuracy of the instrument being compared.

g. Install. The act of emplacing, seating, or fixing into position an item, part, or module (component or assembly) in a manner to allow the proper functioning of the equipment or system.

h. Replace. The act of substituting a serviceable like type part, subassembly, or module (component or assembly) for an unserviceable counterpart.

i. _Repair_. The application of maintenance services (inspect, test, service, adjust, calibrate, replace) or other maintenance actions (welding, grinding, riveting, straightening, facing, remachining, or resurfacing) to restore serviceability to an item by correcting specific damage, fault, malfunction, or failure in a part, subassembly, module (component or assembly) end item, or system.

j. _Overhaul_. That maintenance effort (service/action) necessary to restore an item to a completely serviceable/operational condition as prescribed by maintenance standards (i.e., OMWR) in appropriate technical publications. Overhaul is normally the highest degree of maintenance performed by the Army. Overhaul does not normally return an item to a like-new condition.

k. _Rebuild_. Consists of those services/actions necessary for the restoration-of unserviceable equipment to a like-new condition in accordance with original manufacturing standards. Rebuild is the highest degree of materiel maintenance applied to Army equipment. The rebuild operation includes the act of returning to zero those age measurements (hours, miles, etc.) considered in classifying Army equipments/components.

B-3. Column Entries

a. _Column 1, Group Number_. Column 1 lists group numbers, the purpose of which is to identify components, assemblies, subassemblies, and modules with the next higher assembly.

b. _Column 2, Component/Assembly_. Column 2 contains the noun names of components, assemblies, subassemblies, and modules for which maintenance is authorized.

c. _Column 3, Maintenance Functions_. Column 3 lists the functions to be performed on the item listed in column 2. When items are listed without maintenance functions, it is solely for the purpose of having the group numbers in the MAC and Repair Parts and Special Tools List (RPSTL) coincide.

d. _Column 4, Maintenance Category_. Column 4 specifies, by the listing of a "wore time" figure in the appropriate subcolumn(s), the lowest level of maintenance authorized to perform the function listed in column 3. This figure represents the active time required to perform that maintenance function at the indicated category of maintenance. If the number or complexity of the tasks within the listed maintenance function varies at different maintenance categories, appropriate "work time" figures will be shown for each category. The time required to restore an item (assembly, subassembly, component, module, end item, or system) to a serviceable condition under typical field operating conditions. This time includes preparation time, troubleshooting time, and quality assurance/ quality control time in addition to the time required to perform the specific tasks identified for the maintenance functions authorized in the MAC. Subcolumns of column 4 are as follows:

    O - Organizational
    F - Direct support
    H - General Support
    D - Depot

e. Column 5, Tools and Equipment. Column 5 specifies, by code, those common tool sets (not individual tools), special tools, and test and support equipment required to perform the designated function.

f. Column 6, Remarks. This column contains an alphabetical code which leads to the remark in Section IV, Remarks, which is pertinent to the item opposite the particular code.

B-4. Tool and Test Equipment Requirements (Section III)

a. Tool or Test Equipment Reference Code. The numbers in this column coincide-with the numbers used in the tools and equipment column of the MAC. The numbers indicate the applicable tool or test equipment for the maintenance functions.

b. Maintenance Category. The codes in this column indicate the maintenance category allocated the tool or test equipment.

c. Nomenclature. This column lists the noun name and nomenclature of the tools and test equipment required to perform the maintenance functions.

d. National/North Atlantic Treaty Organization (NATO) Stock Number. This column lists the National/NATO stock number of the specific tool or test equipment.

e. Tool Number. This column lists the manufacturer's part number of the tool followed by the 5-digit Commercial and Government Entity (CAGE) in parentheses.

B-5. Remarks (Section IV)

a. Reference Code. This code refers to the appropriate item in section II, column 6.

b. Remarks. This column provides the required information necessary to clarify items appearing in section II.

B-3

Section II. MAINTENANCE ALLOCATION CHART
FOR
POWER PLANTS AN/MJQ-32/33

(1) GROUP NUMBER	(2) COMPONENT/ ASSEMBLY	(3) MAINT. FUNCTION	(4) MAINTENANCE LEVEL					(5) TOOLS AND EQUIP	(6) REMARKS
			C	O	F	H	D		
01	GENERATOR SET	INSPECT	0.1						A
		TEST		1.0					
		REPLACE			3.0			1	
02	SWITCH BOX	INSPECT	0.1						
		TEST		0.5				2	
		REPLACE		0.5				1	
		REPAIR		0.4	2.0			1, 5	
	SWITCH	INSPECT		0.2					
		TEST		0.5				2	
		REPLACE		1.0				1	
	POST TERMINALS/BOARD	INSPECT		0.1					
		REPLACE		0.5				1	
	WIRING	INSPECT		0.2					
		TEST		0.5				2	
		REPLACE		1.0				1	
		REPAIR			1.0				
	LIGHT AND WIRE ASSEMBLY	INSPECT		0.2					
		TEST		0.2				2	
		REPLACE		0.4				1	
		REPAIR		0.4					
	CABLES	INSPECT		0.1					
		TEST		0.2				2	
		REPLACE		0.2				1	
		REPAIR			1.0				
03	ACCESSORY BOX	INSPECT			1.0				B
		REPLACE	0.1	0.5				1	
		REPAIR			2.0			1	
04	RACK ASSEMBLY, STOWAGE	INSPECT							C
		REPLACE	0.1	0.4				1	
		REPAIR		0.5				1	D

SECTION II. MAINTENANCE ALLOCATION CHART
FOR
POWER PLANTS AN/MJQ-32/33

(1)	(2)	(3)	(4)					(5)	(6)
			MAINTENANCE LEVEL					TOOLS AND	
GROUP NUMBER	COMPONENT/ ASSEMBLY	MAINT. FUNCTION	C	O	F	H	D	EQUIP	REMARKS
05	BRACKETS/HOLERS/SUPPORTS	INSPECT	0.1						
		REPLACE		0.5				1	
	AND GROUND STUD	REPAIR		0.2				3, 4	E
06	TRAILER ASSEMBLY								F
		INSPECT	0.1						
		REPLACE			4.0			1	
	BED/FENDERS	INSPECT	0.1						
		REPLACE			4.0			1	
		REPAIR			4.0			1	G
	LEG PROP	INSPECT	0.1						
		SERVICE		0.2					
		REPLACE			0.5			1	
		REPAIR			0.7			1	
	LIGHTING	INSPECT	0.1						
		TEST			0.3			2	
		REPLACE			1.0			1	
		REPAIR			0.5				
	TARPAULIN SUPPORT AND	INSPECT	0.1					1	B
		REPLACE	0.5						
	BOW ASSEMBLY TARPAULIN	INSPECT	0.1						
		REPLACE	0.5						
		REPAIR			1.0				H

B-5

SECTION III. TOOL AND TEST EQUIPMENT REQUIREMENTS

Tool or Test Equipment Ref Code	Maintenance Category	Nomenclature	National/NATO Stock Number	Tool Number
1	0	Tool Kit, General Mechanics	5180-00-177-7033	
2	0	Multimeter, AN/PSN-45	6625-01-139-2512	
3	0	Riveter, Blind Head	5120-00-148-5847	
4	0	Crimp Tool NS25441	5130-00-762-9100	
5	0	Soldering Gun GT7A-3	3439-00-004-0915	
6	0, F	Drill, 1/4-inch	5130-00-807-3009	

SECTION IV. REMARKS

Reference Code	Remarks
A	See TM5-6115-615-12 for generator set maintenance.
B	AN/MJQ-33 only.
C	AN/MJQ-32 only.
D	Repair by replacement of clamps, runners, and strap fasteners.
E	Repair cable-reel bracket by replacing straps.
F	See TM9-2330-202-14&P for trailer maintenance.
G	Fender/fender extension for AN/MJQ-32 only.
H	Repair in accordance with FM10-16, Fabric Repairing.

B-7/(B-8 blank)

**OPERATOR, UNIT, DIRECT SUPPORT
AND GENERAL SUPPORT MAINTENANCE
REPAIR PARTS AND SPECIAL TOOLS LIST FOR
AN/MJQ-32 AND AN/MJQ-33 POWER PLANTS**

SECTION I. INTRODUCTION

C-1. **SCOPE**. This RPSTL lists and authorizes spares and repair parts; special tools; special test, measurement, and diagnostic equipment (TMDE); and other special support equipment required for performance of operator, unit, and intermediate direct support and general support maintenance of the AN/MJQ32 and AN/MJQ-33 Power Plants. It authorizes the requisitioning, issue, and disposition of spares, repair parts and special tools as indicated by the source, maintenance and recoverability (SMR) codes.

C-2. **GENERAL**. In addition to this section, Introduction, this Repair Parts and Special Tools List is divided into the following sections:

a. **Section II. Repair Parts List**. A list of spares and repair parts authorized by this RPSTL for use in the performance of maintenance. The list also includes parts which must be removed for replacement of the authorized parts. Parts lists are composed of functional groups in ascending alphanumeric sequence, with the parts in each group listed in ascending figure and item number sequence. Bulk materials are listed in item name sequence. Repair parts kits are listed separately in their own functional group within Section II. Repair parts for repairable special tools are also listed in this section. Items listed are shown on the associated illustration(s)figure(s).

b. **Section III. Special Tools List**. A list of special tools, special TMDE, and other special support equipment authorized by this RPSTL (as indicated by Basis of Issue (BOI) information in DESCRIPTION AND USABLE ON CODE column) for the performance of maintenance.

c. **Section IV. Cross-references Indexes**. A list, in National Item Identification Number (NIIN) sequence, of all National stock numbered items appearing in the listing, followed by a list in alphanumeric sequence of all part numbers appearing in the listings. National stock numbers and part numbers are cross-referenced to each illustration figure and item number appearance. The figure and item number index lists figure and item number in alphanumeric sequence and cross- references NSN, FSCM and part number.

C-3. **EXPLANATION OF COLUMNS (SECTIONS II AND III)**.

a. **ITEM NO. (Column (1))**. Indicates the number used to identify items called out in the illustration.

b. **SMR Code (Column (2))**. The Source, Maintenance, and Recoverability (SMR) code is a 5-position code containing supply/requisitioning information, maintenance category authorization criteria, and disposition instruction, as shown in the following breakout:

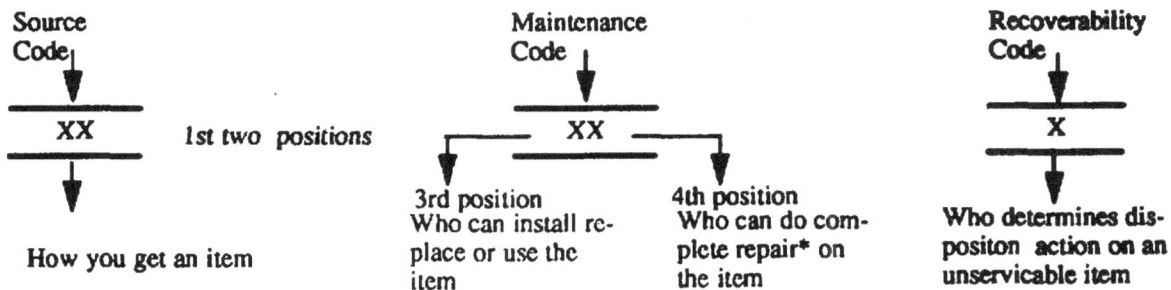

Source Code	Maintenance Code	Recoverability Code
XX 1st two positions	**XX**	**X**
How you get an item	3rd position — Who can install replace or use the item 4th position — Who can do complete repair* on the item	Who determines disposition action on an unservicable item

NOTE

***Complete Repair: Maintenance capacity, capability, and authority to perform all corrective mainte-
nance tasks of the "Repair" function In a use/user environment in order to restore serviceability to a
failed item.**

(1) Source Code. The source code tells you how to get an item needed for maintenance, repair, or overhaul of an end item/equipment. Explanations of source codes follows:

Code

Explanation **PA**
Stocked   **PB**   items; use the applicable NSN to request/requisition items with these source codes. They
are autho-   **PC****   rized to the category indicated by the code entered in the 3d position of the SMR code.
       PD
NOTE: **PE   Items coded PC are subject to deterioration.
       PF
Items   **PG**   with these codes are not to be requested/requisitioned individually. They are par of a kit
which is   authorized to the maintenance category indicated in the 3d position of the SMR code. The
complete kit   must be requisitioned and applied.

       KD
       KF
       KB

Ex-   **AO – – (Assembled by**   planation
      org/AVUM
      Level)   Items with these codes are not to be requested/requisitioned individually. They
must   **AF – – (Assembled by** be made from bulk material which is identified by the part number in the DESCRIP-
      DS/AVIM Level TION and USABLE ON CODE (UOC) column and listed in the Bulk Material group
of   **AH – – (Assembled by** the repair parts list in this RPSTL. f the item is authorized to you by the 3d position
code   **GS Category)**   of the SMR code, but the source code indicates it is made at a higher level, order
the   **AL – – (Assembled by** item from the higher level of maintenance.
      SRA)
   AD – – (Assembled by
      Depot)

```
AO -- (Assembled by
      org/AVUM
      Level)
AF -- (Assembled by
      DS/AVIM Level
AH -- (Assembled by
      GS Category)
AL -- (Assembled by
      SRA)
AD -- (Assembled by
      Depot)
```

Explanation

items with these codes are not to be requested/requisitioned individually. The parts that make up the assembled item must be requisitioned or fabricated and assembled at the level of maintenance indicated by the source code. If the 3d position code of the SMR code authorizes you to replace the item, but the source code indicates the items are assembled at a higher level, order the item from the higher level of maintenance.

Code Explanation

XA - - Do not requisition "XA" -coded item. Order its next higher assembly. (Also, refer to the NOTE below.)
XB - - If an "XB" item is not available from salvage, order it using the FSCM and part number given.
XC - - Installation drawing, diagram, instruction sheet, field service drawing, that is identified by manufacturer's part number.
XD - - Item is not stocked. Order an "XD" -coded item through normal supply channels using the FSCM and part number given, if no NSN is available.

NOTE

Cannibalization or controlled exchange, when authorized, may be used as a source of supply for items with the above source codes, except for those source coded "XA" or those aircraft support items restricted by requirements of AR 750-1.

(2) **Maintenance Code.** Maintenance codes tells you the level(s) of maintenance authorized to USE and REPAIR support items. The maintenance codes are entered in the third and fourth positions of the SMR code as follows:

(a) The maintenance code entered in the third position tells you the lowest maintenance level authorized to remove, replace, and use an item. The maintenance code entered in the third position will indicate authorization to one of the following levels of maintenance.

Code Application/Explanation

C - Crew or operator maintenance done within organizational or aviation unit maintenance.
O - Organizational or aviation unit category can remove, replace, and use the item.

F - Direct support or aviation intermediate level can remove, replace, and use the item.
H - General support level can remove, replace, and use the item.
L - Specialized repair activity can remove, replace, and use the item.
D - Depot level can remove, replace, and use the item.

    (b) The maintenance code entered in the fourth position tells whether or not the item is to be repaired and identifies the lowest maintenance level with the capability to do complete repair (i.e., perform all authorized repair functions.)

NOTE
Some limited repair may be done on the item at a lower level of maintenance, if authorized by the Maintenance Allocation Chart (MAC) and SMR codes. This position will contain one of the following maintenance codes.

Code Application/Explanation
O - Organizational or (aviation unit) is the lowest level that can do complete repair of the item.
F - Direct support or aviation intermediate is the lowest level that can do complete repair of the item.
H - General Support is the lowest level that can do complete repair of the item.
L - Specialized repair activity is the lowest level that can do complete repair of the item.
D - Depot is the lowest level that can do complete repair of the item.
Z - Nonreparable. No repair is authorized.
B - No repair is authorized. (No parts or special tools are authorized for the maintenance of a "B" coded item). However, the item may be reconditioned by adjusting, lubricating, etc., at the user level.

    (3) **Recoverability Code.** Recoverability codes are assigned to items to indicate the disposition action on unserviceable items. The recoverability code is entered in the fifth position of the SMR Code as follows:

Recoverability
    Codes Application/Explanation
Z - Nonreparable item. When unserviceable, condemn and dispose of the item at the level of maintenance shown in 3d position of SMR Code.

Recoverability
Codes Application/Explanation

O - Reparable item. When uneconomically reparable, condemn and dispose of the item at organizational or aviation
unit level
F - Reparable item. When uneconomically reparable, condemn and dispose of the item at the direct support or
aviation
intermediate level
H - Reparable item. When uneconomically reparable, condemn and dispose of the item at the general support level.
D - Reparable item. When beyond lower level repair capability, return to depot. Condemnation and disposal of item
not authorized below depot level.
L - Reparable item. Condemnation and disposal not authorized below specialized repair activity (SRA).
A -Item requires special handling or condemnation procedures because of specific reasons (e.g., precious metal content, high dollar value, critical material, or hazardous material). Refer to appropriate manuals/directives for specific
instructions.

c. **FSCM (Column (3)).** The Federal Supply Code for Manufacturer (FSCM) is a 5-digit numeric code which is used to
identify the manufacturer, distributor, or Government agency, etc., that supplies the item.

d. **PART NUMBER (Column (4)).** Indicates the primary number used by the manufacturer, (individual, company, firm,
corporation, or Government activity), which controls the design and characteristics of the item by means of its engineering
drawings, specifications standards, and inspection requirements to identify an item or range of items.

NOTE
**When you use an NSN to requisition an item, the item you receive may have a different part number
from the part ordered.**

e. DESCRIPTION AND USABLE ON CODE (UOC) (Column (5)). This column includes the following information:

(1) The Federal item name and, when required, a minimum description to identify the item.
(2) The physical security classification of the item is indicated by the parenthetical entry, e.g., Phy Sec Cl-
Confidential, Phy Sec Cl (S) - Secret, Phy Sec Cl (T) - Top Secret.
(3) Items that are included in kits and sets are listed below the name of the kit or set.
(4) Spare/repair parts that make up an assembled item are listed immediately following the assembled item
line entry.
(5) Part numbers for bulk materials are referenced in this column in the line item entry for the item to be
manufactured fabricated.
(6) When the item is not used with all serial numbers of the same model, the effective serial numbers are
shown on the last line(s) of the description (before UOC).

C-5

(7) The usable on code, when applicable (see paragraph 5, Special Information).

(8) In the Special Tools List section, the basis of issue (BOI) appears as the last line(s) In the entry for each special tool, special TMDE, and other special support equipment. When density of equipments supported exceeds density spread indicated in the basis of issue the total authorization is increased proportionately.

(9) The statement "END OF FIGURE' appears just below the last item description in Column 5 for a given figure in both Section II and Section III.

(10) The indenture, shown as dots appearing before the repair part, indicates that the item is a repair part of the next higher assembly.

f. OTY (Column (6)). The QTY (quantity per figure column) indicates the quantity of the item used in the breakout shown on the illustration figure, which is prepared for a functional group, subfunctional group, or an assembly. A "V" appearing in this column in lieu of a quantity indicates that the quantity is variable and may vary from application to application.

C-4. EXPLANATION OF COLUMNS (SECTION IV).

a. NATIONAL STOCK NUMBER (NSN) INDEX.

(1) STOCK NUMBER column. This column lists the NSN by National item identification number (NIIN) sequence. The NIIN consists of the last nine digits of the
When using this column to locate an item, ignore the first 4 digits of the NSN. However, the complete NSN should be used

$$\text{NSN, i.e. } (5305-01-574-1467). \quad \frac{\text{NSN}}{\text{NIIN}}$$

when ordering items by stock number.

(2) FIG. column. This column lists the number of the figure where the item is identified/located. The figures are in numerical order in Section II and Section III.

(3) ITEM column. The item number identifies the item associated with the figure listed in the adjacent FIG. column. This item is also identified by the NSN listed on the same line.

b. PART NUMBER INDEX. Part numbers in this index are listed by part number in ascending alphanumeric sequence (i.e. vertical arrangement of letter and number combination which places the first letter or digit of each group in order A through Z, followed by the numbers 0 through 9 and each following letter or digit in like order).

(1) FSCM column. The Federal Supply Code for Manufacturer (FSCM) Is a 5digit numeric code used to identify the manufacturer, distributor, or Government agency, etc., that supplies the item.

(2) PART NUMBER column. Indicates the primary number used by the manufacturer (individual, firm, corporation, or Government activity), which controls the design and characteristics of the item by means of its engineering drawings, specifications standards, and inspection requirements to identify an item or range of items.

(3) STOCK NUMBER column. This column lists the NSN for the associated part number and manufacturer identified in the PART NUMBER and FSCM columns to the left.

(4) FIG. column. This column lists the number of the figure where the item is identified/located in Sections II and III.

(5) ITEM column. The item number is that number assigned to the item as it appears in the figure referenced in adjacent figure number column.

c. FIGURE AND ITEM NUMBER INDEX.

(1) FIG. column. This column lists the number of the figure where the item is identified/located in Section II and III.

(2) ITEM column. The item number is that number assigned to the item as it appears in the figure referenced in the adjacent figure number column.

(3) STOCK NUMBER column. This column lists the NSN for the item.

(4) FSCM column. The Federal Supply Code for Manufacturer (FSCM) is a 5-digit numeric code used to identify the manufacturer, distributor, or Government agency, etc., that supplies the item.

(5) PART NUMBER column. Indicates the primary number used by the manufacturer (individual, firm, corporation, or Government activity), which controls the design and characteristics of the item by means of its engineering drawings, specifications standards, and inspection requirements to identify an item or range of items.

C-5. SPECIAL INFORMATION.

a. **USABLE ON CODE.** Identification of the usable codes of the publication for various DOD models are:

Code Used On

EPH AN/MJQ-32
EPJ AN/MJQ-33

b. **PUBLICATIONS.** Publications pertaining to items comprising AN/MJQ-32 and AN/MJQ-33 are located in appendix A..

C-6. HOW TO LOCATE REPAIR PARTS.

a. When National Stock Number or Part Number Is NOT known.

(1) First. Using the table of contents, determine the assembly group or subassembly group to which the item belongs. This is necessary since figures are prepared for assembly groups and subassembly groups, and listings are divided into the same groups.

(2) Second. Find the figure covering the assembly group or subassembly group to which the item belongs.

(3) Third. Identify the item on the figure and note the item number.

(4) Fourth. Refer to the Repair Parts List for the figure to find the part number for the item number noted on the figure.

(5) Fifth. Refer to the Part Number Index to find the NSN, if assigned.

b. When National Stock Number or Part Number Is Known:

(1) First. Using the Index of National Stock Numbers and Part Numbers, find the pertinent National Stock Number or Part Number. The NSN index is in National Item Identification Number (NIIN) sequence (see c-4a.(1)). The part numbers in the Part Number index are listed in ascending alphanumeric sequence (see paragraph c-4.b). Both indexes
cross-reference you to the illustration figure and item number of the item you are looking for.

(2) Second. After finding the figure and item number, verify that the item is the one you are looking for, then locate the item number in the repair parts list for the figure.

7. ABBREVIATIONS. Not applicable.

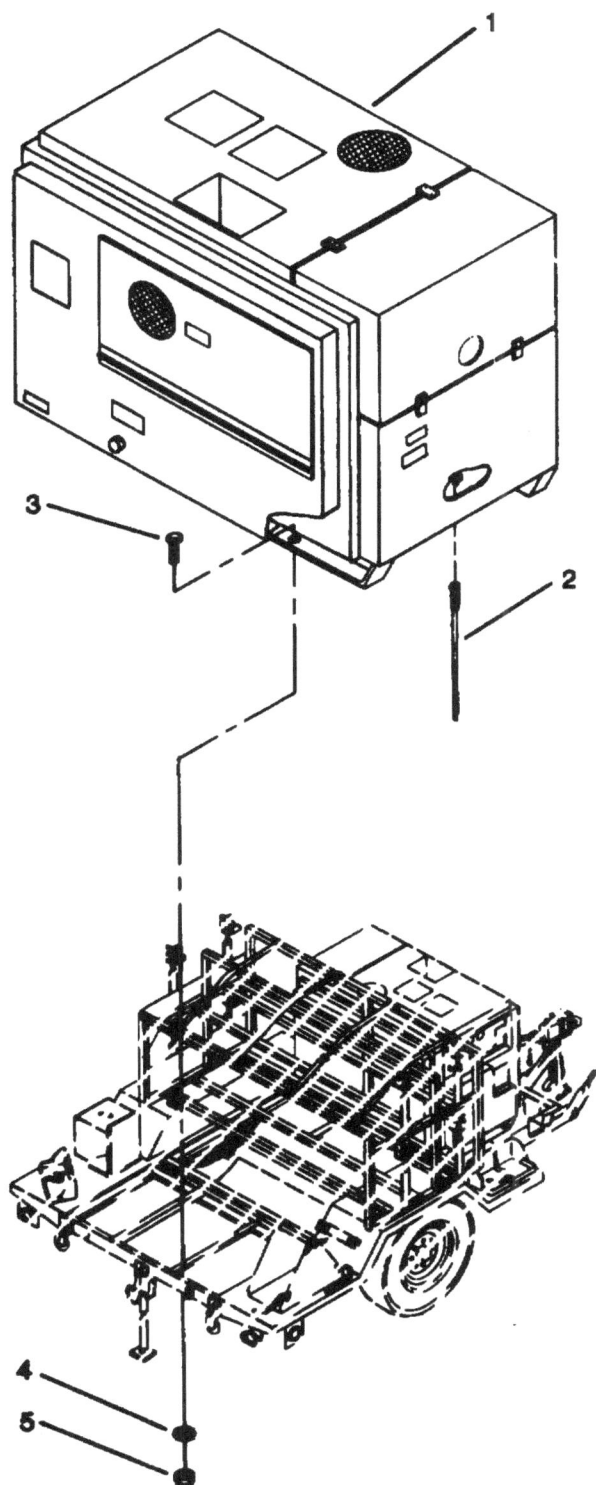

Figure C-1. Generator Set.

ITEM SMR PART
 NO CODE FSCM NUMBER DESCRIPTION AND USABLE ON CODES(UOC) QTY

GROUP 01 GENERATOR SET

FIG. C-1 GENERATOR SET

1 PDOFF 30554 MEP-701A GENERATOR SET,DIESE 3 KW 60 HZ/................................ 1
 (ASK) ....
2 PAOZZ 97403 13229E2304 HOSE,FUEL DRAIN... 1
3 PAOZZ 96906 MS90728-111 SCREWDCAP,HEXAGON H 8
4 PAOZZ 96906 MS27183-18 WASHER,FLAT... 4
5 PAOZZ 96906 MS51922-33 NUT,SELF-LOCKING,HE.. 8

END OF FIGURE

C-11

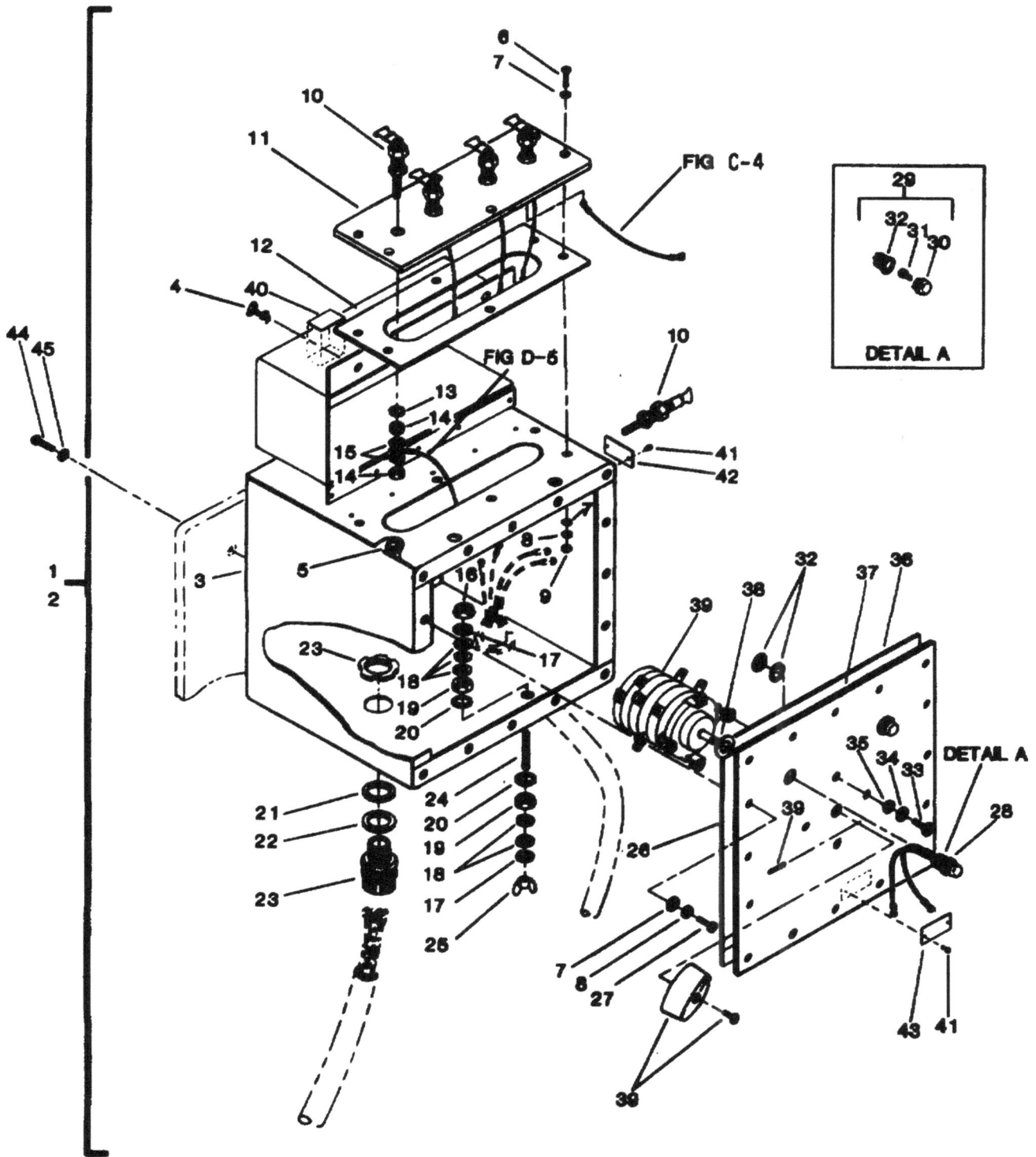

Figure C-2. Five-Wire Switch Box.

(1) (2) (3) (4) (5) (6)

ITEM SMR PART
 NO CODE FSCM NUMBER DESCRIPTION AND USABLE ON CODES(UOC) QTY

GROUP 02 SWITCH BOX

FIG. C-2 FIVE-WIRE SWITCH BOX

1 PAOFH 97403 13205E5079-3 DISTRIBUTION BOX		1
UOC:EPH		
2 PAOZZ 97403 13205E5079-4 DISTRIBUTION BOX		1
UOC:EPJ		
3 XBFFF 97403 13216E7600 .BOX		1
4 PAOZZ 81349 MIL-F-5591 ..FASTNER, WING HD TYPE 3..............................		2
5 PAOZZ 96906 MS20427-4C6 ..RIVET, SOLID		4
6 PAOZZ 96906 MS35206-267 .SCREW, MACHINE #10-24 X 1.000" LG.........................		6
7 PAOZZ 96906 MS27183-8 .WASHER, FLAT #10 REG STEEL	28	
8 PAOZZ 96906 MS35338-43 .WASHER, LOCK #10 REG STEEL	22	
9 PAOZZ 96906 MS35649-202 .NUT, PLAIN, HEXAGON #10-24 STEEL		6
10 PAOZZ 96906 MS39347-2 .TERMINAL, STUD SERVICE AND GROUND		5
11 PBFZZ 97403 13212E3606 .TERMINAL BOARD		1
12 PAOZZ 97403 13212E3610 .GASKET		1
13 PAOZZ 96906 MS35333-108 .WASHER, LOCK		5
14 PAOZZ 96906 MS16203-37 .NUT, PLAIN, HEXAGON	10	
15 PAOZZ 96906 MS35338-101 .WASHER, LOCK 1/4" PHB	10	
16 PAOZZ 96906 MS16203-27 .NUT, PLAIN, HEXAGON CLASS B TYPE 2.....................		1
17 PAOZZ 96906 MS35338-103 .WASHER, LOCK 3/8 REG PHB		2
18 PAOZZ 88044 AN961-616T .WASHER, FLAT		5
19 PAOZZ 96906 MS16203-39 .NUT, PLAIN, HEXAGON		2
20 PAOZZ 96906 MS35333-110 .WASHER, LOCK		2
21 XBFZZ 97403 13218E5140-1 .GASKET		2
22 PAOZZ 97403 13218E5139-1 .WASHER, FLAT...		2
23 PAOZZ 97403 13218E5149-2 .STUFFING TUBE		2
24 PAOZZ 97403 13214E1223 .STUD, CONTINUOUS THR		1
25 PAOZZ 96906 MS35425-28 .NUT, PLAIN, WING 3/8-16 UNC CRS		1
26 MOOZZ 81349 MIL-R-6130 .RUBBER, CELLULAR MAKE FROM MIL-R-		2
6130, 9.5 IN LG		
27 PAOZZ 96906 MS35207-263 .SCREW, MACHINE *10-32 X .500", ...	16	
STEEL		
28 PAOOZ 97403 13212E3560 .LIGHT, INDICATOR		2
29 PAOOZ 97403 13214E1391 ..LIGHT, INDICATOR		2
30 PAOZZ 72619 181-0937-003 ..LENS, CLEAR...................................		2
31 PAOZZ 58224 NE2G ..LAMP		2
32 PAOZZ 72619 181-8836-09-553 ..HOUSING..		2
33 PAOZZ 96906 MS35206-281 .SCREW, MACHINE 1/4-20 X .750", ...		4
STEEL		
34 PAOZZ 96906 MS27183-11 .WASHER, FLAT		4
35 PAOZZ 80205 NAS1598-4Y .WASHER, SEALING 0.25" ID X .5 OD		4
36 MOOZZ 81349 MIL-R-6130 .RUBBER, CELLULAR MAKE FROM MIL-R-		1
6130, 10.5 in. LONG.		
37 PAOZZ 97403 13205E5078 .COVER		1
38 XBFZZ 97403 13218E5160 .SEAL, ROTARY SW ITCH.............................		1
39 PBOFF 97403 13219E9860 .SWITCH, ROTARY		1
40 MDOZZ 97403 13216E7603 .PLATE WARNING		1
41 PAOZZ 96906 MS21318-20 .SCREW, DRIVE *4 X .188" STEEL....................		1
42 MDOZZ 97403 13226E5889-1 .PLATE, INFORMATION		1

(1) (2) (3) (4) (5) (6)

ITEM SMR PART
NO CODE FSCM NUMBER DESCRIPTION AND USABLE ON CODES(UOC) QTY

43 PAOZZ 97403 13226E5889-2 .PLATE INFORMATION	1
44 PAOZZ 96906 NS90728-60 SCREW, CAP, HEXAGON H	4
45 PAOZZ 96906 NS27183-57 WASHER, FLAT	4

END OF FIGURE

C-14

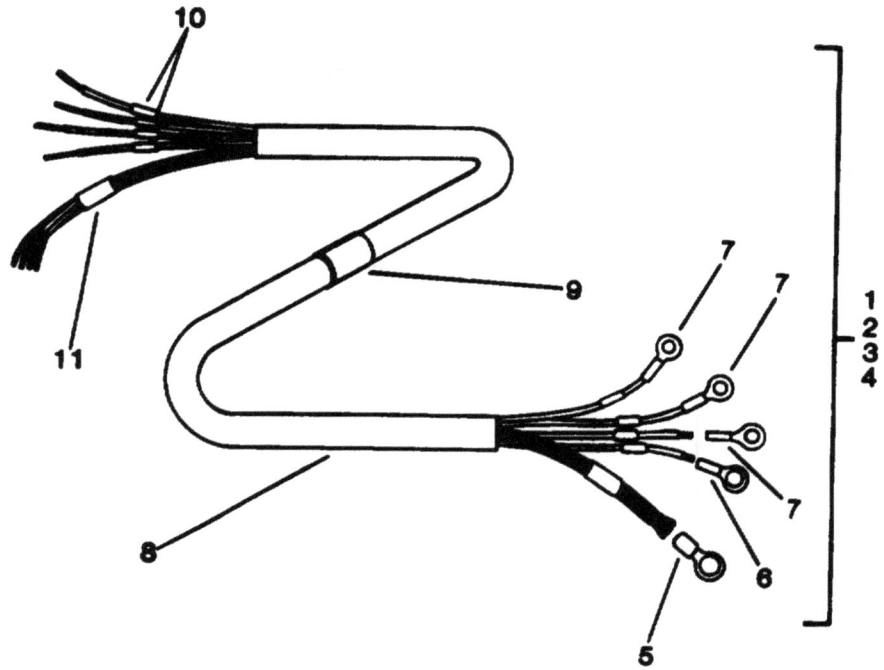

Figure C-3. Five-Wire Power Cable.

C-15/(C-16 Blank)

(1) (2) (3) (4) (5) (6)

ITEM SMR PART
NO CODE FSCM NUMBER DESCRIPTION AND USABLE ON CODES(UOC) QTY

GROUP 02 SWITCH BOX

FIG. C-3 FIVE-WIRE POWER CABLE

1 PAOFF 97403 13212E3571-4 .CABLE LENGTH,112.00 IN	1
UOC:EPH	
2 PAOFF 97403 13212E3570-3 .CABLE LENGTH,148.00 IN	1
UOC:EPH	
3 PAOFF 97403 13212E3571-5 .CABLE LENGTH,112.00 IN	1
UOC:EPJ	
4 PAOZZ 97403 13212E3570-4 .CABLE LENGTH,72.00 IN	1
UOC:EPJ	
5 PAOZZ 96906 MS25036-114 ..TERMINAL,LUG 12-10 RI 3/8" YELLOW	1
6 PAOZZ 96906 MS25036-157 ..TERMINAL,LUG 12-10 RI 1/4" YELLOW	1
7 PAOZZ 96906 MS25036-112 ..TERMINAL,LUG 12-10 RI * 10 YELLOW	3 8
PAOZZ 81349 CO-04HDE ..CABLE, ELECTRIC 600V, HVY DTY	V
9 PAOZZ 81349 M23053/5-110-9 ..INSULATION SLEEVING HEAT SHRINK	1
10 MFFZZ 81349 M23053/5-107-9 ..INSULATION SLEEVING MAKE FROM..............................	4
M23053/5-107-9,1.0 IN LG..	
11 PAOZZ 81348 QQW343C06BIB ..WIRE,ELECTRICAL	V
UOC:EPH	

END OF FIGURE

C-17

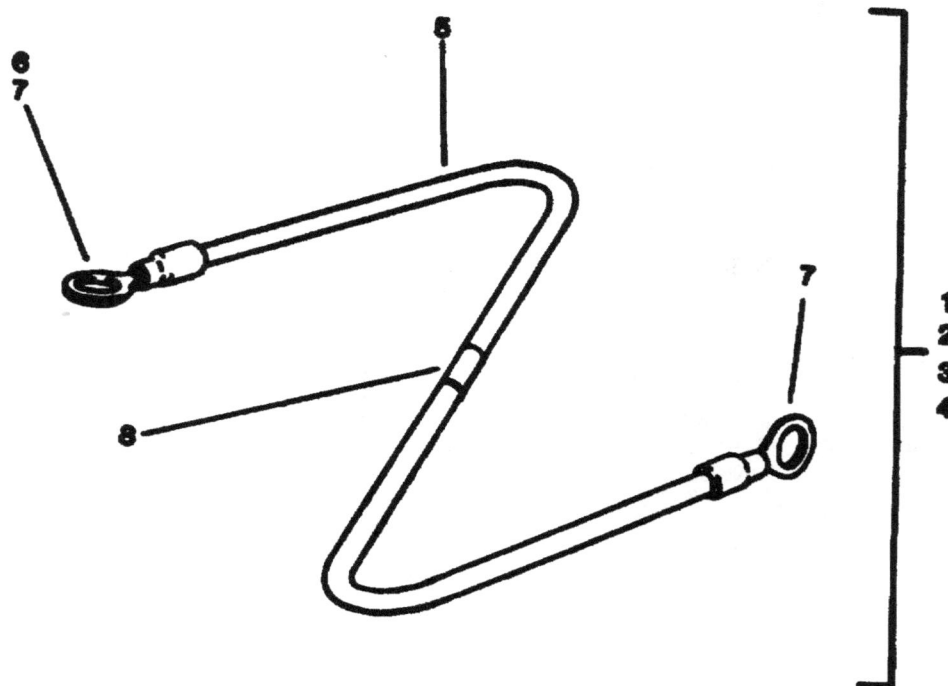

Figure C-4. Switch Box Electrical Leads.

(1) (2) (3) (4) (5) (6)

ITEM SMR PART
NO CODE FSCM NUMBER DESCRIPTION AND USABLE ON CODES(UOC) QTY

GROUP 02 SWITCH BOX

FIG. C-4 SWITCH BOX ELECTRICAL LEADS

1 PAOFF 97403 13212E3567-1 .LEAD,ELECTRICAL USE LUGS 6&7 ON W3..................... 1
2 PAOFF 97403 13212E3567-2 .LEAD,ELECTRICAL USE LUGS 6&7 ON W4.................... 1
3 PAOFF 97403 13212E3567-3 .LEAD,ELECTRICAL USE LUGS 6&7 ON W5.................... 1
4 PAOFF 97403 13212E3567-4 .LEAD,ELECTRICAL USE LUG #7 ON............................. 1
          BOTH ENDS OF W6 ....
5 PAOZZ 81349 M5086/2-10 ..CABLE,ELECTRIC 600V, 1OAWG, HVY V
          DTY ....
6 PAOZZ 96906 MS25036-112 ..TERMINAL,LUG USE ON W3,W4,W5,12-......................... 3
          10 #10 YELLOW ....
7 PAOZZ 96906 MS25036-157 ..TERMINAL,LUG USE ON W6,12-10 10............................ 2
          YELLOW ....
8 PAOZZ 81349 M23053-15-105-5 ..INSULATION SLEEVE 1

END OF FIGURE

C-19

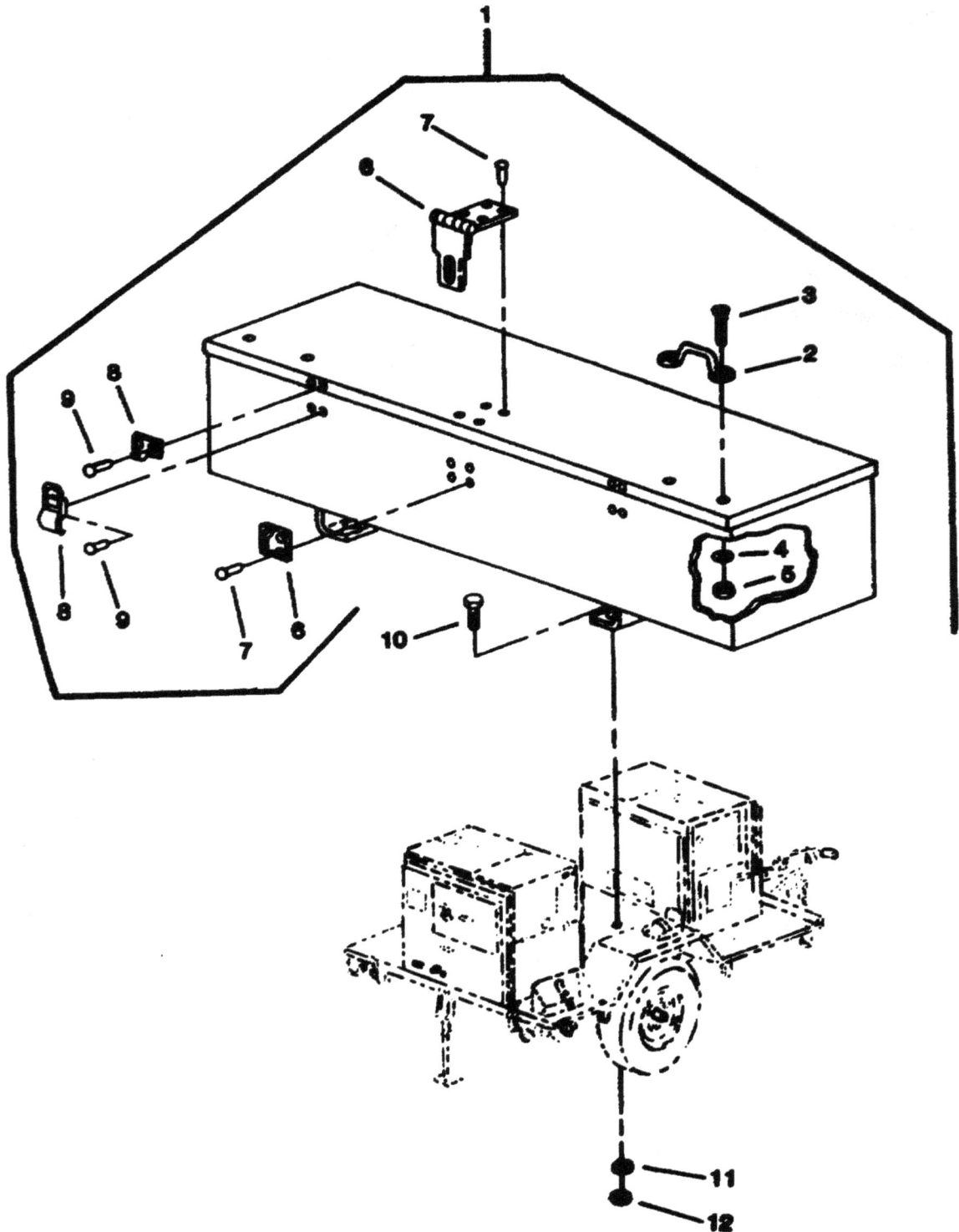

Figure C-5. Accessory Box, AN/MJQ-33.

(1) (2) (3) (4) (5) (6)

ITEM SMR PART
NO CODE FSCM NUMBER DESCRIPTION AND USABLE ON CODES(UOC) QTY

GROUP 03 ACCESSORY BOX

FIG. C-5 ACCESSORY BOX, AN/MJQ-33

1 XBOFO 97403 13226E7737 CHEST,ACCESSORY	1
UOC:EPJ	
2 PAOZZ 96906 MS51939-3 .LOOP,STRAP FASTENER STRAP	2
UOC:EPJ	
3 PAOZZ 96906 MS24693-S273 .SCREW,MACHINE	4
UOC:EPJ	
4 PAOZZ 96906 MS27183-42 .WASHER,FLAT o10 WIDE STEEL	4
UOC:EPJ	
5 PAOZZ 96906 MS21046C3 .NUT,SELF-LOCKING,HE	4
UOC:EPJ	
6 PAOZZ 96906 MS27969-4 .HASP,HINGED	8
UOC:EPJ	
7 PAOZZ 96906 MS9460-102 .RIVET,SOLID	8
UOC:EPJ	
8 PAOZZ 96906 MS18015-1 .CATCH,CLAMPING	8
UOC:EPJ	
9 PAOZZ 96906 MS20613-4P5 .RIVET,SOLID FASTNER,TY2,CL3 16	
UOCTEPJ	
10 PAOZZ 96906 MS90728-32 BOLT,MACHINE 5/16-18 UNC-2A X .750	4
"	
UOC:EPJ	
11 PAOZZ 96906 MS27183-56 WASHER,FLAT .375 ID NOM.............................. 4	
UOC:EPJ	
12 PAOZZ 96906 MS51922-9 NUT,SELF-LOCKING,HE 5/16-18 UNC-ZB......................... 4	
UOC:EPJ	

END OF FIGURE

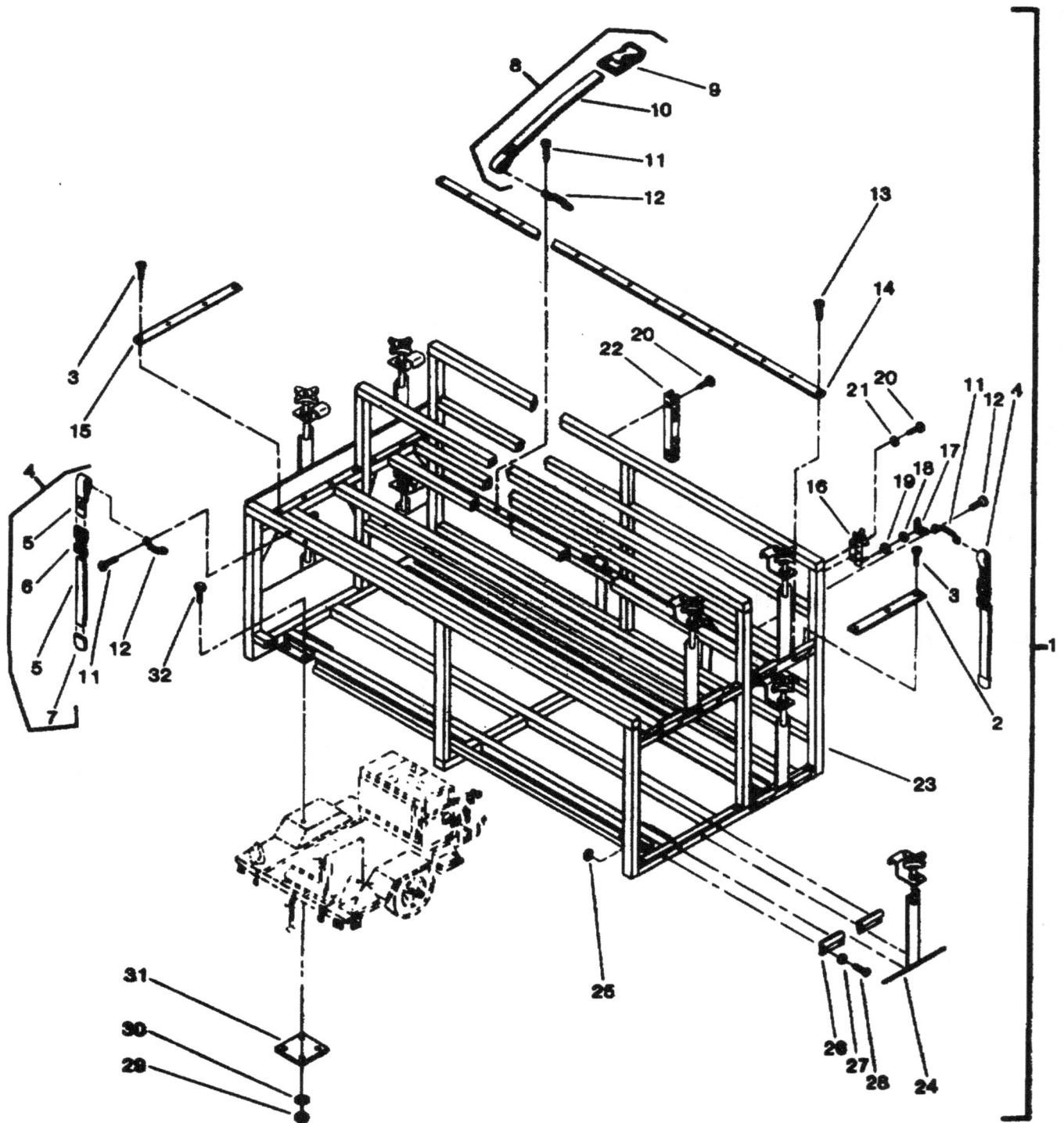

Figure C-6. Stowage Rack Assembly, AN/INQ-32.

(1) (2) (3) (4) (5) (6)

ITEM SMR PART
NO CODE FSCM NUMBER DESCRIPTION AND USABLE ON CODES(UOC) QTY

GROUP 04 RACK ASSEMBLY, STOWAGE

FIG. C-6 STOWAGE RACK ASSEMBLY,
AN/MJQ-32

1 PBOOO 97403 13228E9902 RACK,ASSY STOWAGE			1
UOC:EPH			
2 XBOZZ 97403 13205E5120 .CLAMP,RUNNER			4
UOC:EPH			
3 PAOZZ 96906 MS51960-67 .SCREW,MACHINE .190-32UNF-2A X 28			
.750", CSK 82..........................			
UOC:EPH			
4 MFOFF 97403 13218E5091 .TIEDOWN,STRAP MAKE FROM MIL-W-			2
4088,36.25 IN LG			
UOC:EPH			
5 PAOZZ 81349 MIL-W-4088 ..WEBBING TEXTILE WOVEN NYLON, OD			2
Y7..			
UOC:EPH			
6 PAOZZ 96906 MS51929-2 ..BUCKLE SIZE 1, CS, CAD OR ZINC			2
PLATED			
UOC:EPH			
7 PAOZZ 96906 MS51926-3 ..CLIP,END,STRAP END STRAP			2
UOC:EPH			
8 MFOFF 97403 13216E7504 .STRAP WEBBING			4
UOCIEPH			
9 PAOZZ 96906 MS51929-2 ..BUCKLE SIZE 1, CS, CAD OR ZINC			3
PLATED			
UOC:EPH			
10 PAOZZ 81349 MIL-W-4088 ..WEBBING TEXTILE WOVEN NYLON, OD			3
Y77			
UOC:EPH			
11 PAOZZ 96906 MS24628-24 .SCREW,SELF-TAPPING 20			
UOC:EPH			
12 PAOZZ 96906 MS51939-3 .LOOP,STRAP FASTENER LOOP STRAP 10			
UOC:EPH			
13 PAOZZ 96906 MS24628-24 .SCREW,SELF-TAPPING...........................			1
UOC:EPH			
14 XBOZZ 97403 13205E5123 .RUNNER..			8
UOC:EPH			
15 XBOZZ 97403 13205E5121 .CLAMP,RUNNER			4
UOC:EPH			
16 PAOOZ 97403 13212E3617 .CARRI ER,ROD,GROUND			1
UOC:EPH			
17 PAOZZ 96906 MS35425-70 NUT,PLAIN,WING 1/4-20 UNC STEEL.................			2
UOC:EPH			
18 PAOZZ 96906 MS35338-44 WASHER,LOCK 1/4 REG STEEL			2
UOC:EPH			
19 PAOZZ 96906 MS27183-10 WASHER,FLAT 1/4 REG STEEL			2
UOC:EPH			
20 PAOZZ 96906 MS51957-81 .SCREW,MACHINE 1/4-20 UNC X .750"			4
CRS			
UOC:EPH			

(1) (2) (3) (4) (5) (6)

ITEM SMR PART
NO CODE FSCM NUMBER DESCRIPTION AND USABLE ON CODES(UOC) QTY

21 PAOZZ 96906 MS15795-810 .WASHER,FLAT 1/4 REG CRS		4
UOC:EPH		
22 XBOZZ 97403 13228E9899 .BRACKET GROUND RODS		1
UOC:EPH		
23 XAOFF 97403 13228E9906 .RACK,STOWAE		1
UOC:EPH		
24 XBOZZ 97403 13205E5137-2 .CLAMP,SCREW,QUICK A		8
UOC:EPH		
25 PAOZZ 96906 MS51922-1 .NUT,SELF-LOCKING,HE 1/4-20 UNC		1
STEEL		
UOC:EPH		
26 PAOZZ 97403 13205E5125 .LEAF,BUTT HINGE 16		
UOC:EPH		
27 PAOZZ 96906 NS27183-52 .WASHER,FLAT .281 NOM ID 18		
UOC:EPH		
28 PAOZZ 96906 NS90728-13 .SCREW,CAP,HEXAGON H 1/4-20 UNC-2A		32
X .750" GR8		
UOC:EPH		
29 PAOZZ 96906 NS51922-17 NUT,SELF-LOCKING,HE 16		
UOC:EPH		
30 PAOZZ 96906 MS27183-57 WASHER,FLAT 16		
UOC:EPH		
31 XBOZZ 97403 13228E9907 PLATE,BACKING		4
UOC:EPH		
32 PAOZZ 96906 MS90728-60 SCREW,CAP,HEXAGON H 16		
UOC:EPH		

END OF FIGURE

C-24

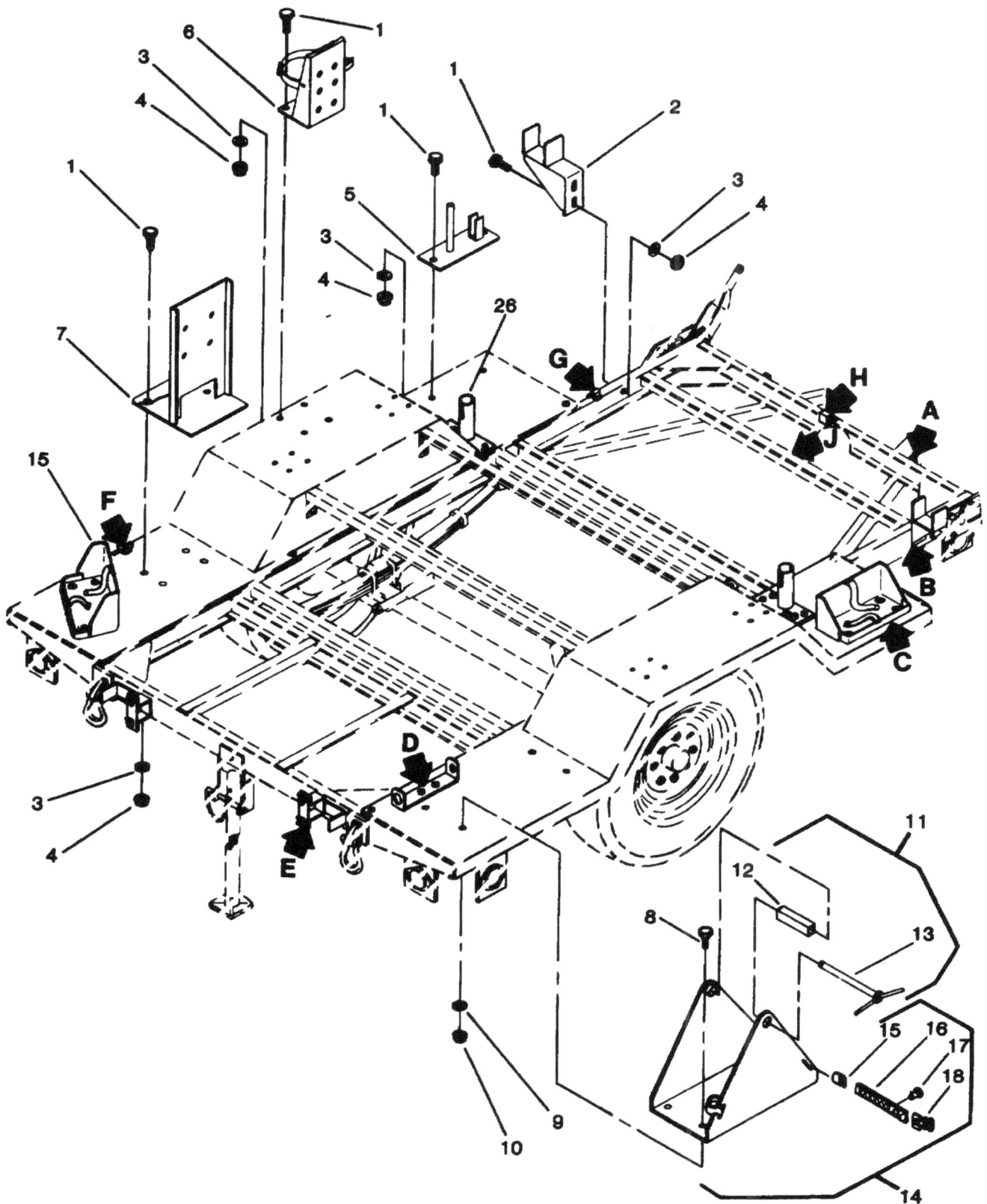

Figure C-7. Brackets, Holders, Supports, and Ground Studs, AN/MJQ-32 (Sheet 1 of 3).

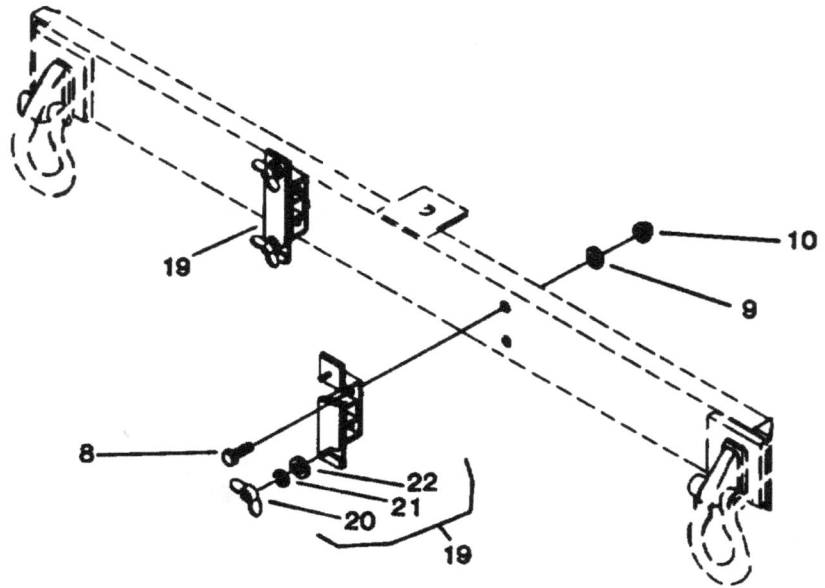

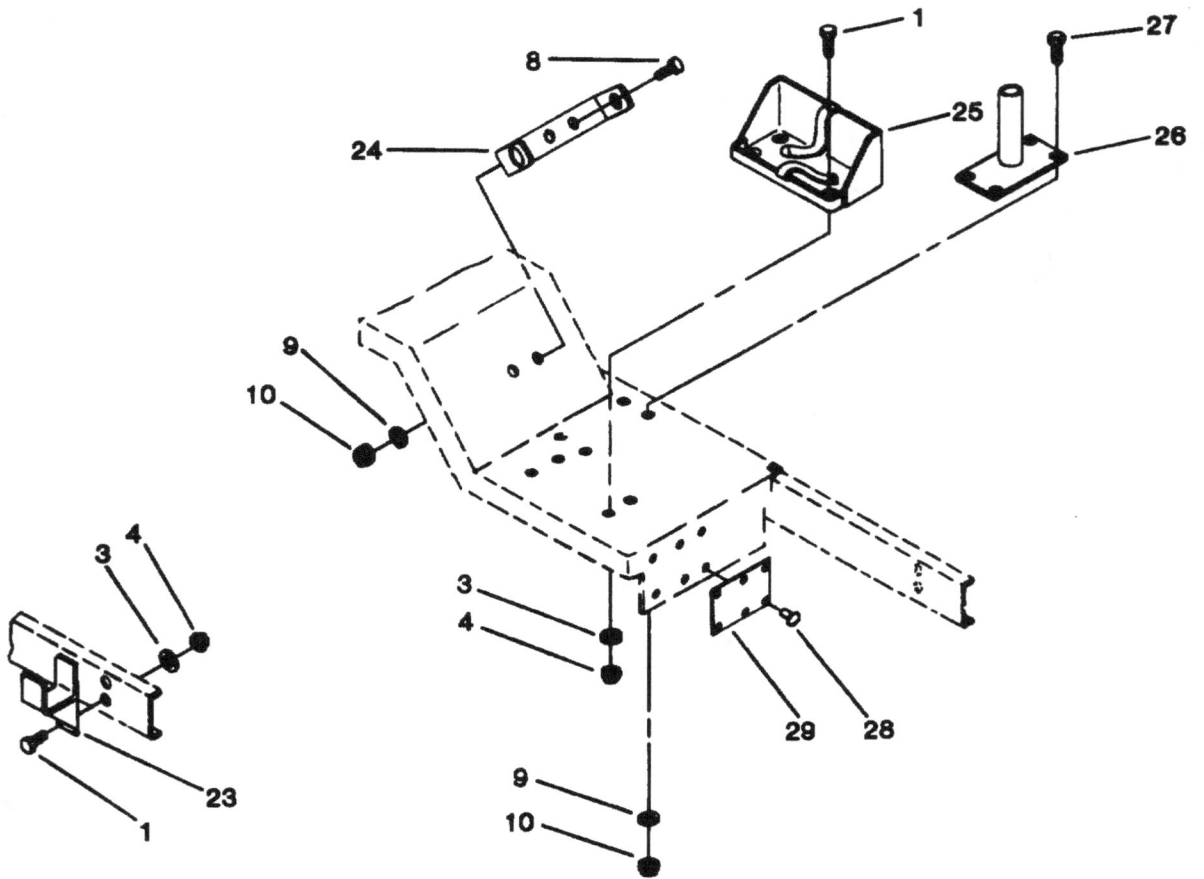

Figure C-7. Brackets, Holders, Supports, and Ground Studs,
AN/MJQ-32 (Sheet 2 of 3).

DETAIL D

DETAIL E

DETAIL F

DETAIL G

DETAIL H

DETAIL J

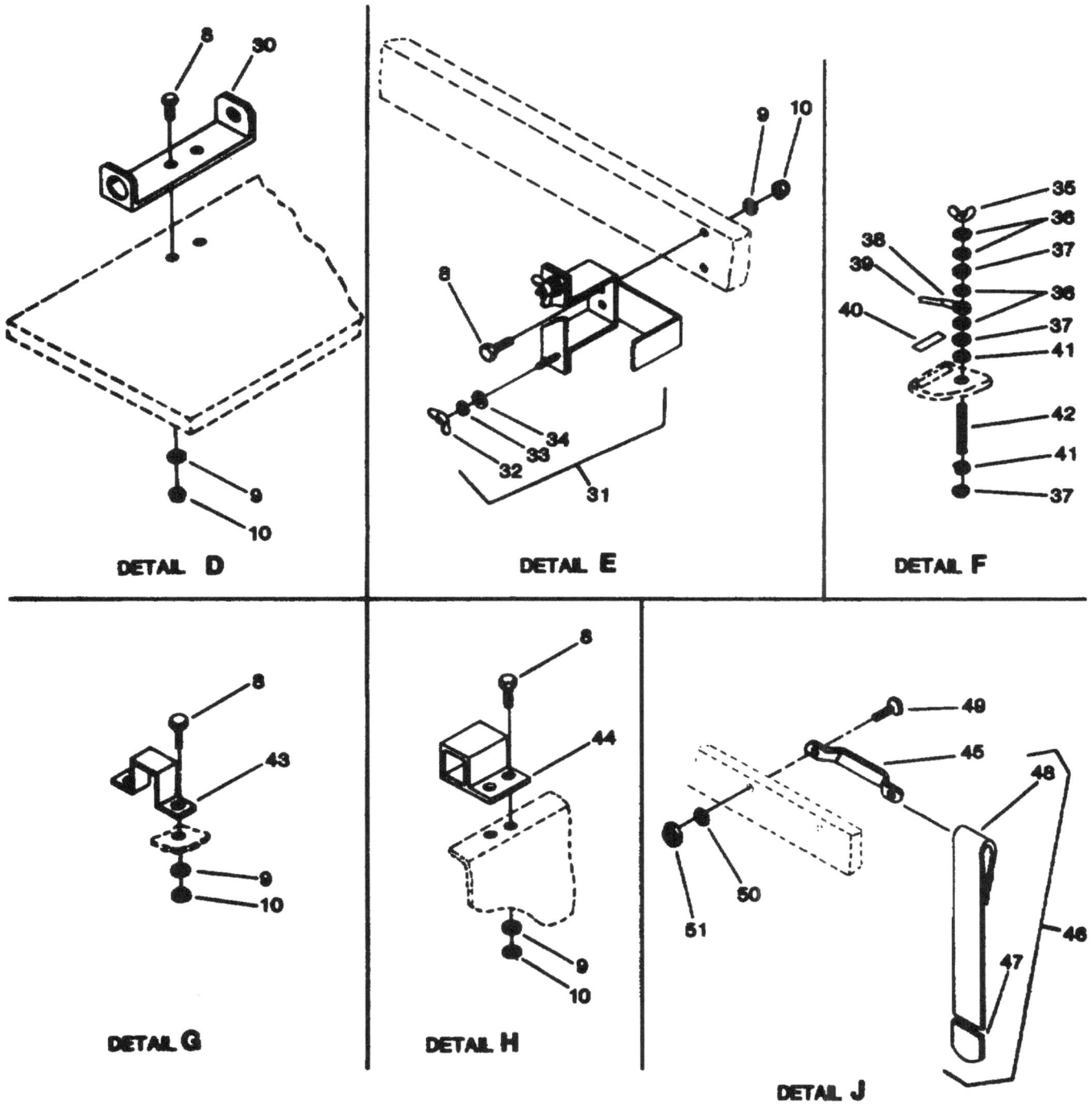

Figure C-7. Brackets, Holders , Supports, and Ground Studs,
AN/1JQ-32 (Sheet 3 of 3).

(1) (2) (3) (4) (5) (6)

ITEM SMR PART
 NO CODE FSCM NUMBER DESCRIPTION AND USABLE ON CODES(UOC) QTY

GROUP 05 BRACKETS/HOLDERS/SUPPORTS
AND GROUND STUD

FIG. C-7 BRACKETS,HOLDERS,SUPPORTS
AND GROUND STUDS,AN/NMJQ-32

1 PAOZZ 96906 MS90728-60 SCREW,CAP,HEXAGON H 3/8-16 UNC-2A   1
                    X 1.00" ....
                    UOC:EPH
2 XBOZZ 97403 13228E9897-2 MAST SUPPORT..   1
                    UOC:EPH
3 PAOZZ 96906 MS27183-57 WASHER,FLAT .406 NOM ID...................... 18
4 PAOZZ 96906 MS51922-17 NUT,SELF-LOCKING,HE 3/8-16 UNC 18
5 XBOZZ 97403 13228E9898 PLATE MTG,DRIVER/PU..   1
                    UOC:EPH
6 PAOZZ 97403 13214E1235 BRAKE, FIRE EXTING   1
                    UOC:EPH
7 PAOZZ 97403 13229E2303-2 BRKT,MTG DIST BOX...   1
                    UOC:EPH
8 PAOZZ 96906 MS90728-6 SCREW,CAP,HEXAGON H 1/4-20 UNC X 13
                    .750", GR8. ....
                    UOC:EPH
9 PAOZZ 96906 MS27183-52 WASHER,FLAT .281 NOM ID 21
                    UOCTEPH
10 PAOZZ 96906 MS51922-1 NUT,SELF-LOCKING,HE 1/4-20 UNC   21
                    STEEL ....
                    UOC:EPH
11 XBOOO 97403 13216E7605 HOLD-DOWN ASSY. CAB...   1
                    UOC:EPH
12 XBOZZ 97403 13216E7607 .SPINDLE,REEL...   1
                    UOC:EPH
13 XBOZZ 97403 13216E7606-1 .HOLD-DOWN REEL...   1
                    UOC:EPH
14 XBOOO 97403 13217E2062 BRACKET,CABLE REELL   1
                    UOC:EPH
15 PAOZZ 96906 MS51926-3 .CLIP,END,STRAP END STRAP   1
                    UOC:EPH
16 MFOFF 81349 MIL-W-530 .WEBBING,TEXTILE COTTON, TY2, CL4.............................   1
                    UOC:EPH
17 PAOZZ 96906 MS9319-208 .RIVET,SOLID   1
                    UOC:EPH
18 PAOZZ 96906 MS51929-2 .BUCKLE   2
                    UOC:EPH
19 PAOOZ 97403 13212E3617 CARRIER,ROD,GROUND   2
                    UOC:EPH
20 PAOZZ 96906 MS35425-70 .NUT,PLAIN,WING 1/4-20 UNC STEEL..............................   4
                    UOC EPH
21 PAOZZ 96906 MS35338-44 .WASHER,LOCK 1/4 REG STEEL   4
                    UOC:EPH
22 PAOZZ 96906 MS27183-10 .WASHER,FLAT 1/4 REG STEEL............................   4
                    UOC:EPH
23 XBOZZ 97403 13228E9897-1 MAST SUPPORT...   1

(1) (2) (3) (4) (5) (6)

ITEM SMR PART
 NO CODE FSCM NUMBER DESCRIPTION AND USABLE ON CODES(UOC) QTY

	UOC:EPH	
24 XBOZZ 97403 13212E3553-2 HOLDER		1
	UOC:EPH	
25 PAOZZ 96906 MS53052-1 BRACKET ASSEMBLY,LI LIQUID CONT		1
	UOC:EPH	
26 XBOZZ 97403 13205E5143 SUPPORT		1
	UOC:EPH	
27 PAOZZ 96906 MS90728-8 SCREW,CAP,HEXAGON H 3/8-16 UNC-2A		4
X 1.25-, GR 6		
	UOC:EPH	
28 PAOZZ 96906 MS21318-20 SCREW,DRIVE *4 X .188" STEEL		6
	UOC:EPH	
29 NDOZZ 97403 13228E6394-21 PU/PP ID/TRANSPORT REFER TO DWG		1
13228E6394-21		
	UOC:EPH	
30 XBOZZ 97403 13212E3553-1 HOLDER,FLEXIBLE SPO		1
	UOC:EPH	
31 PAOOZ 97403 13214E1214 BRACKET,ANGLE		2
	UOC:EPH	
32 PAOZZ 96906 MS35425-70 .NUT,PLAIN,WING 1/4-20 UNC STEEL..........................		2
	UOC:EPH	
33 PAOZZ 96906 MS35338-44 .WASHER,LOCK 1/4 REG STEEL		2
	UOC:EPH	
34 PAOZZ 96906 MS27183-10 .WASHER,FLAT 1/4 REG STEEL		2
	UOC:EPH	
35 PAOZZ 96906 MS35425-28 NUT,PLAIN,WING 3/8-16 UNC CRS		1
	UOC:EPH	
36 PAOZZ 88044 AN961-616T WASHER,FLAT		4
	UOC:EPH	
37 PAOZZ 96906 MS16203-39 NUT,PLAIN.HEXAGON		1
	UOC:EPH	
38 PAOZZ 96906 MS25036-122 TERMINAL,LUG 6 RI 3/8" BLUE		1
	UOC:EPH	
39 PAOZZ 81348 QQW343CO6BIB WIRE,ELECTRICAL		1
	UOC:EPH	
40 MDOZZ 97403 13205E4918 ..PLATE ID GROUND		1
	UOCSEPH	
41 PAOZZ 96906 MS35335-91 WASHER,LOCK		2
	UOC:EPH	
42 PAOZZ 97403 13214E1223 STUD,CONTINUOUS THR		1
	UOC:EPH	
43 PAOZZ 97403 13212E3612 CLAMP,CABLE		3
	UOC:EPH	
44 MDOZZ 97403 SK-M-Q-002-TGM CABLE,BRACKET		3
	UOC:EPH	
45 PAOZZ 96906 MS51939-3 LOOP,STRAP FASTENER		3
	UOC:EPH	
46 MFOFZ 97403 13216E7505-2 STRAP,WEBBING		3
	UOC:EPH	
47 PAOZZ 81349 MIL-C-496 .CLIP		3
	UOC:EPH	
48 MFOFF 81349 MIL-W-4088 .WEBBING TEXTILE		3

(1) (2) (3) (4) (5) (6)

ITEM SMR PART
NO CODE FSCM NUMBER DESCRIPTION AND USABLE ON CODES(UOC) QTY

	UOC:EPH		
49 PAOZZ 96906 MS35207-265 SCREW,MACHINE			6
	UOC:EPH		
50 PAOZZ 96906 MS27183-42 WASHER,FLAT			6
	UOC:EPH		
51 PAOZZ 96906 MS21044-N3 NUT,SELF-LOCKING,HE			6
	UOC:EPH		

END OF FIGURE

C-31

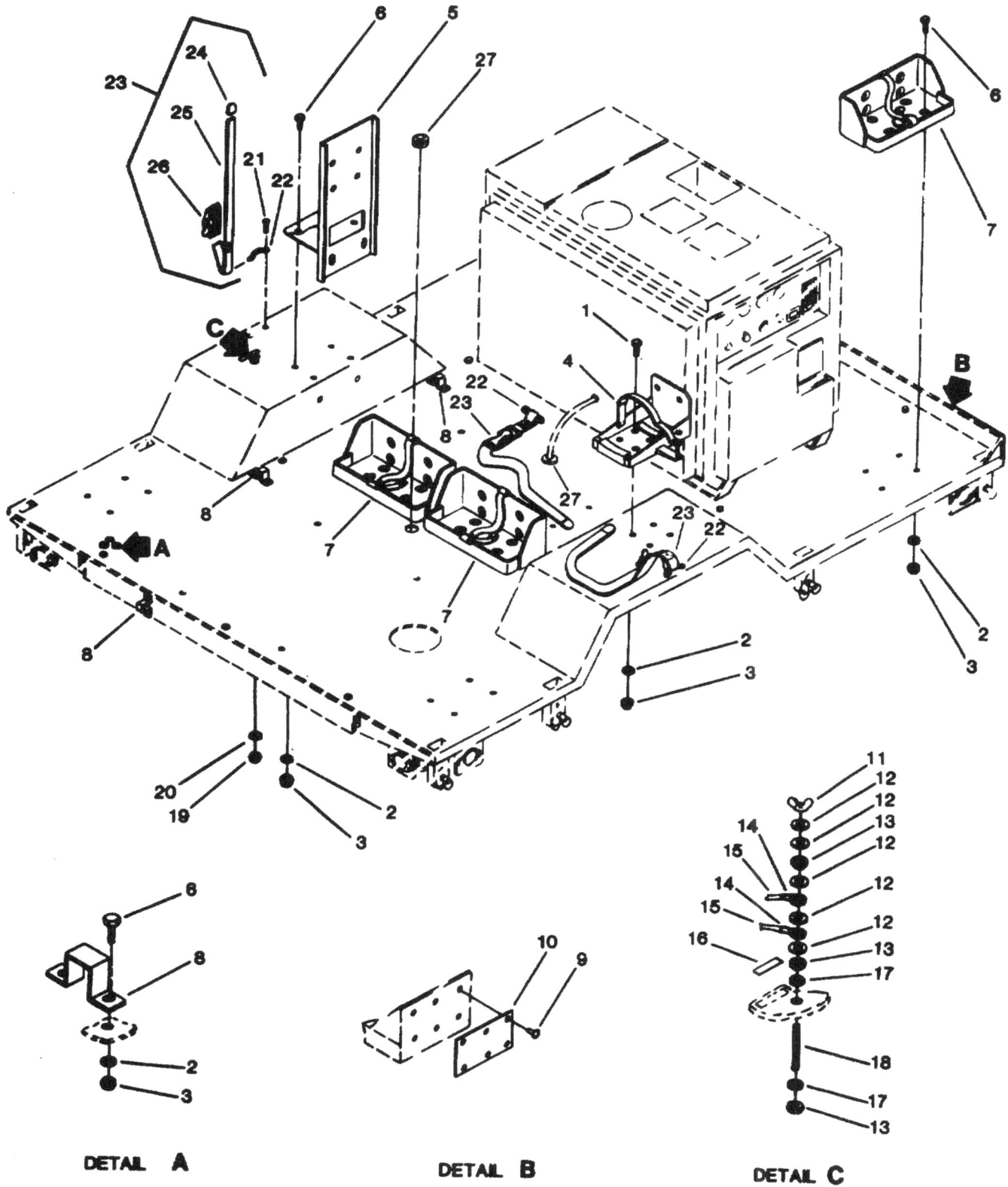

Figure C-8. Brackets, Holders, Supports, and Ground Studs, AN/MJQ-33.

(1) (2) (3) (4) (5) (6)

ITEM SMR PART
NO CODE FSCM NUMBER DESCRIPTION AND USABLE ON CODES(UOC) QTY

GROUP 05 BRACKETS/HOLDERS/SUPPORTS
AND GROUND STUDS

FIG. C-8 BRACKETS,HOLDERS,SUPPORTS
AND GROUND STUDS,AN/MJQ-33

1 PAOZZ 96906 MS90728-58 SCREW,CAP,HEXAGON H		4
UOC:EPJ		
2 PAOZZ 96906 MS27183-57 WASHER,FLAT .406 NOM ID	30	
UOC:EPJ		
3 PAOZZ 96906 MS51922-17 NUT,SELF-LOCKING,HE 3/8-16 UNC	30	
UOC:EPJ		
4 PAOZZ 97403 13214E1235 BRKT,FIRE EXTNG		1
UOC:EPJ		
5 XBOZZ 97403 13229E2303-1 BRKT,MT SWITCH BOX		1
UOC:EPJ		
6 PAOZZ 96906 MS90728-60 SCREW,CAP,HEXAGON H 3/8-16 UNC-2A		22
X 1.00"		
UOC:EPJ		
7 PAOZZ 96906 MS53052-1 BRACKET ASSEMBLY,LI LIQUID CONT		4
UOC:EPJ		
8 PAOZZ 97403 13212E3612 STRAP,RETAINING		5
UOC:EPJ		
9 PAOZZ 96906 MS21318-20 SCREW,DRIVE #4 X .188" STEEL		6
10 MDOZZ 97403 13228E6394-22 DATA P LATE		1
UOC:EPJ		
11 PAOZZ 96906 MS35425-28 NUT,PLAIN,WING 3/8-16 UNC CRS		1
UOC:EPJ		
12 PAOZZ 88044 AN961-616T WASHER,FLAT		4
UOC:EPJ		
13 PAOZZ 96906 MS16203-39 NUT,PLAIN,HEXAGON		3
UOC:EPJ		
14 PAOZZ 96906 MS25036-122 TERMINAL,LUG 6 RI 3/8" BLUE		2
UOC:EPJ		
15 PAOZZ 81348 QQW343CO6B1B WIRE, ELECTRICAL		2
UOC:EPJ		
16 MDOZZ 97403 13205E4918 PLATE ID GROUND		1
UOC:EPJ		
17 PAOZZ 96906 MS35335-91 WASHER,LOCK		2
UOC:EPJ		
18 PAOZZ 97403 13214E1223 STUD,CONTINUOUS THR		1
UOC:EPJ		
19 PAOZZ 96906 MS21044-N3 NUT,SELF-LOCKING,HE		6
UOC:EPJ		
20 PAOZZ 96906 MS27183-42 WASHER,FLAT..................................		6
UOC:EPJ		
21 PAOZZ 96906 MS35207-265 SCREW,MACHINE		6
UOC:EPJ		
22 PAOZZ 96906 MS51939-3 LOOP,STRAP FASTENER		3
UOC:EPJ		
23 MFOFF 97403 13218E5091 TIEDOWN,STRAP		3

(1) (2) (3) (4) (5) (6)

ITEM SMR PART
NO CODE FSCM NUMBER DESCRIPTION AND USABLE ON CODES(UOC) QTY

		UOC:EPJ	
24 PAOZZ 96906 MS51926-3	.CLIP,ENDDSTRAP	....	3
		UOC:EPJ	
25 MFOFF 81349 MIL-W-4088	.WEBBING TEXTILE	....	3
		UOCIEPJ	
26 PAOZZ 96906 MS51929-2	.BUCKLE	....	3
		UOC:EPJ	
27 PAOZZ 96906 MS35489-54	GROMMET,NONMETALLIC	....	2
		UOC:EPJ	

END OF FIGURE

C-34

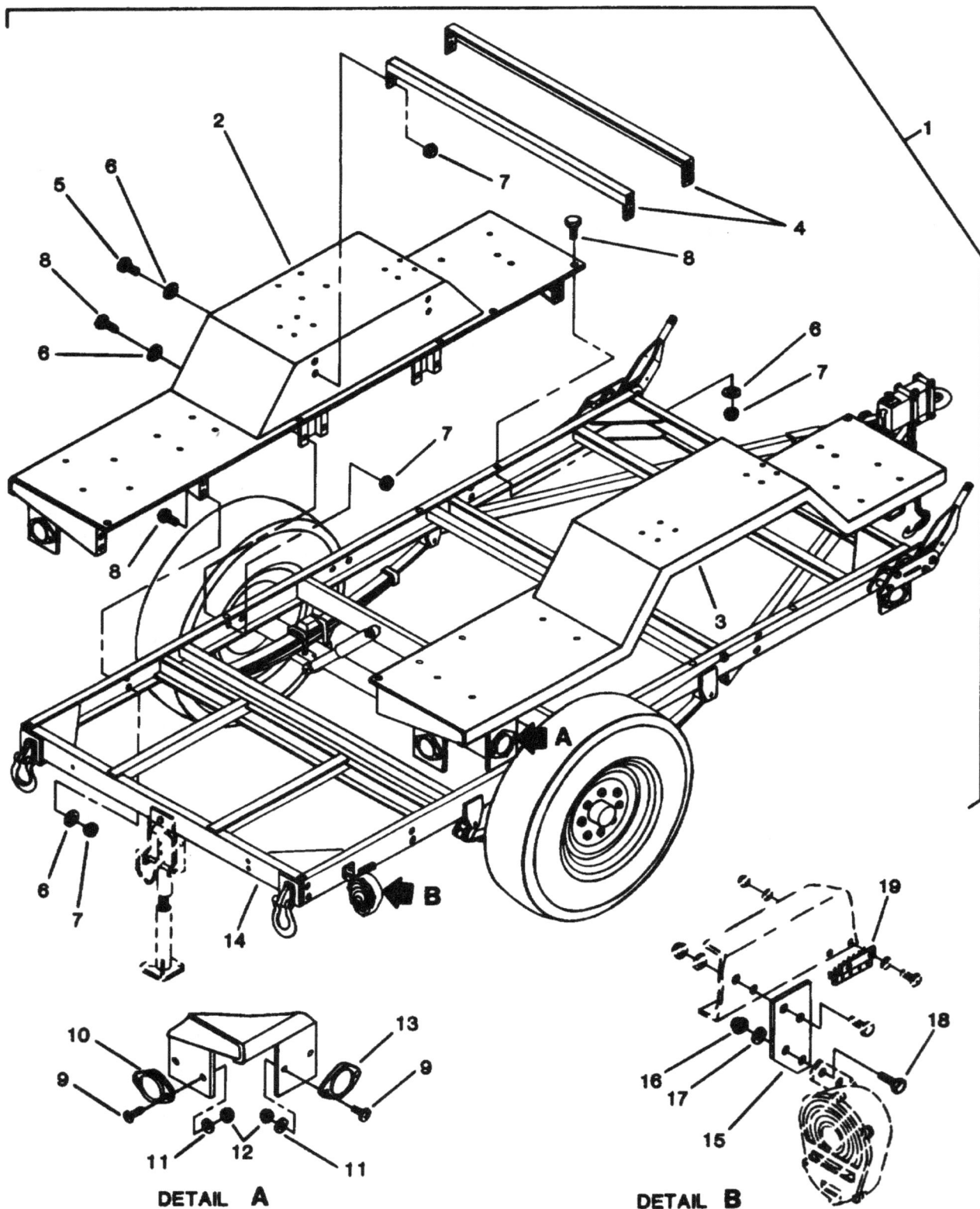

Figure C-9. Trailer Assembly, AN/NJQ-32.

C-35/(C-36 Blank)

(1) (2) (3) (4) (5) (6)

ITEM SMR PART
NO CODE FSCM NUMBER DESCRIPTION AND USABLE ON CODES(UOC) QTY

GROUP 06 TRAILER ASSEMBLY

FIG. C-9 TRAILER ASSEMBLY,AN/MJQ-32

1 PDOFF 97403 13228E9896 TRAILER, 3/4 TON	1
UOC:EPH	
2 XDFFF 97403 13228E9904 .FENDER,RS 3/4 TON C	1
UOC:EPH	
3 XDFFF 97403 13228E9901 .FENDERPCS,3/4 TON C	2
UOC:EPH	
4 XBFZZ 97403 13228E9903 .CROSS BRACE FENDER	2
UOC:EPH	
5 PAOZZ 96906 MS90728-64 .SCREW,CAP,HEXAGON H	8
UOC:EPH	
6 PAOZZ 96906 MS27183-57 .WASHER ,FLAT 10	
UOC:EPH	
7 PAOZZ 96906 MS51922-17 .NUT,SELF-LOCKINGIHE 10	
UOC:EPH	
8 PAOZZ 96906 MS90728-60 .SCREW,CAPPHEXAGON H 10	
UOC:EPH	
9 PAOZZ 96906 MS35206-280 .SCREW,MACHINE 1/4-20 UNC X .625"	12
STEEL	
UOC:EPH	
10 PAOZZ 96906 MS35387-1 .REFLECTOR,INDICATIN RED INDICATING	4
UOC:EPH	
11 PAOZZ 96906 MS27183-52 .WASHER,FLAT .281 NOM ID 12	
UOC:EPH	
12 PAOZZ 96906 MS51922-1 .NUT,SELF-LOCKING,HE 1/4-20 UNC	12
STEEL	
UOC:EPH	
13 PAOZZ 96906 MS35387-2 .REFLECTOR,INDICATIN AMBER	2
INDICATING	
UOC:EPH	
14 PDOFF 97403 13228E9900-2 .CHASSIS,TRAILER MOD	1
UOC:EPH	
15 XBOZZ 97403 13228E9905 ..PLATE,TAILLIGHT REL...	2
UOC:EPH	
16 PAOZZ 96906 MS51922-17 ..NUT,SELF-LOCKING,HE 3/8-16 UNC..................................	4
UOC:EPH	
17 PAOZZ 96906 MS27183-57 ..WASHER,FLAT .406 NOM ID	4
UOC:EPH	
18 PAOZZ 96906 MS90728-60 ..SCREW,CAP,HEXAGON H 3/8-16 UNC-	4
2A X 1.00"..	
UOC:EPH	
19 PAOZZ 19207 8747908-1 ..CLIP ASSY	2
UOC:EPH	

END OF FIGURE

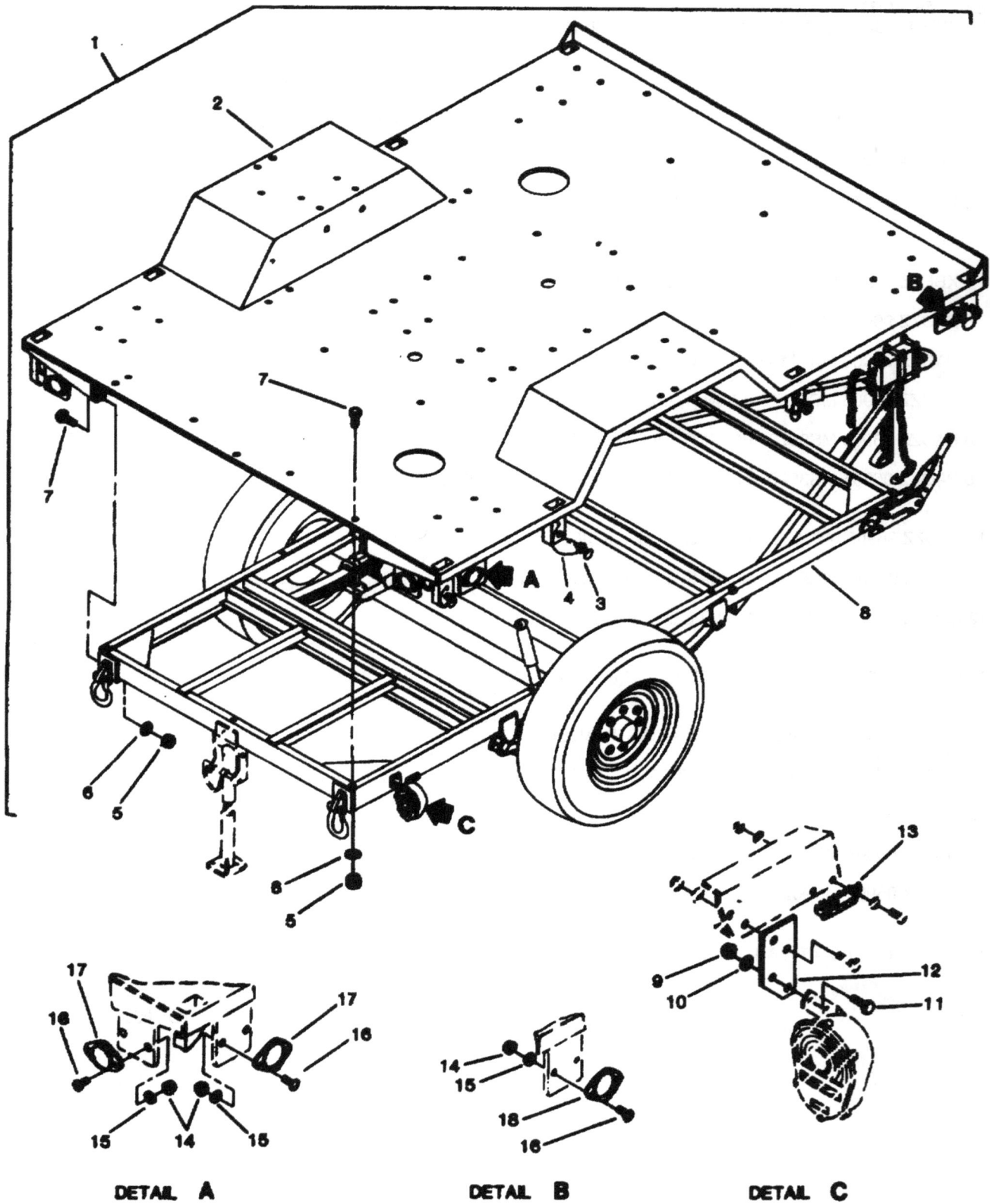

DETAIL A

DETAIL B

DETAIL C

Figure C-10. Trailer Assembly, AN/MJQ-33.

ITEM SMR PART
NO CODE FSCM NUMBER DESCRIPTION AND USABLE ON CODES(UOC) QTY

GROUP 06 TRAILER ASSEMBLY

FIG. C-10 TRAILER ASSEMBLY, AN/MJQ-33

1 PDOFF 97403 13229E2302 TRAILER ASSEMBLY 3/		1
UOC:EPJ		
2 XBFZZ 97403 13221E7326 .BODY, TRAILER		1
UOC:EPJ		
3 XBFZZ 96906 MS17990-C613 PIN,QUICK RELEASE POSTIVE LOCKING		8
UOC:EPJ		
4 XBFZZ 81348 RR-C-271 TY2CL7 CHAIN,WELDLESS TYPE 2 CLASS 7...................................		8
UOC:EPJ		
5 PAOZZ 96906 MS51922-17 .NUT,SELF-LOCKING,HE 3/8-16 UNC 29		
UOC:EPJ		
6 PAOZZ 96906 MS27183-57 .WASHERPFLAT .406 NOM ID 32		
UOC:EPJ		
7 PAOZZ 96906 MS90728-65 .SCREW,CAP,HEXAGON H 3/8-16 UNC-2A		26
X 1.75" GR8...		
UOC:EPJ		
8 PDOFF 97403 13228E9900-1 .CHASSIS 3/4 TON MOD		1
UOC:EPJ		
9 PAOZZ 96906 MS51922-17 ..NUT,SELF-LOCKING,HE 3/8-16 UNC.............................		4
UOC:EPJ		
10 PAOZZ 96906 MS27183-57 ..WASHER,FLAT .406 NOM ID..		4
UOC:EPJ		
11 PAOZZ 96906 MS90728-60 ..SCREW,CAP,HEXAGON H 3/8-16 UNC-		4
2A X 1.00"		
UOC:EPJ		
12 PAOZZ 97403 13228E9905 ..PLATE,TAILLIGHT REL...		2
UOC:EPJ		
13 PAOZZ 19207 8747908-1 ..CLIP ASSY		2
UOC:EPJ		
14 PAOZZ 96906 MS51922-1 .NUT,SELF-LOCKING,HE 1/4-20 UNC		12
STEEL		
UOC:EPJ		
15 PAOZZ 96906 MS27183-52 .WASHER,FLAT .281 NOM ID. 12		
UOC:EPJ		
16 PAOZZ 96906 MS90728-3 .SCREW,CAP,HEXAGON H 1/4-20 UNC-2A		12
X .500" GR8		
UOC:EPJ		
17 PAOZZ 96906 MS35387-1 .REFLECTOR,INDICATIN RED INDICATING		4
UOC:EPJ		
18 PAOZZ 96906 MS35387-2 .REFLECTOR,INDICATIN AMBER		2
INDICATING		
UOC:EPJ		

END OF FIGURE

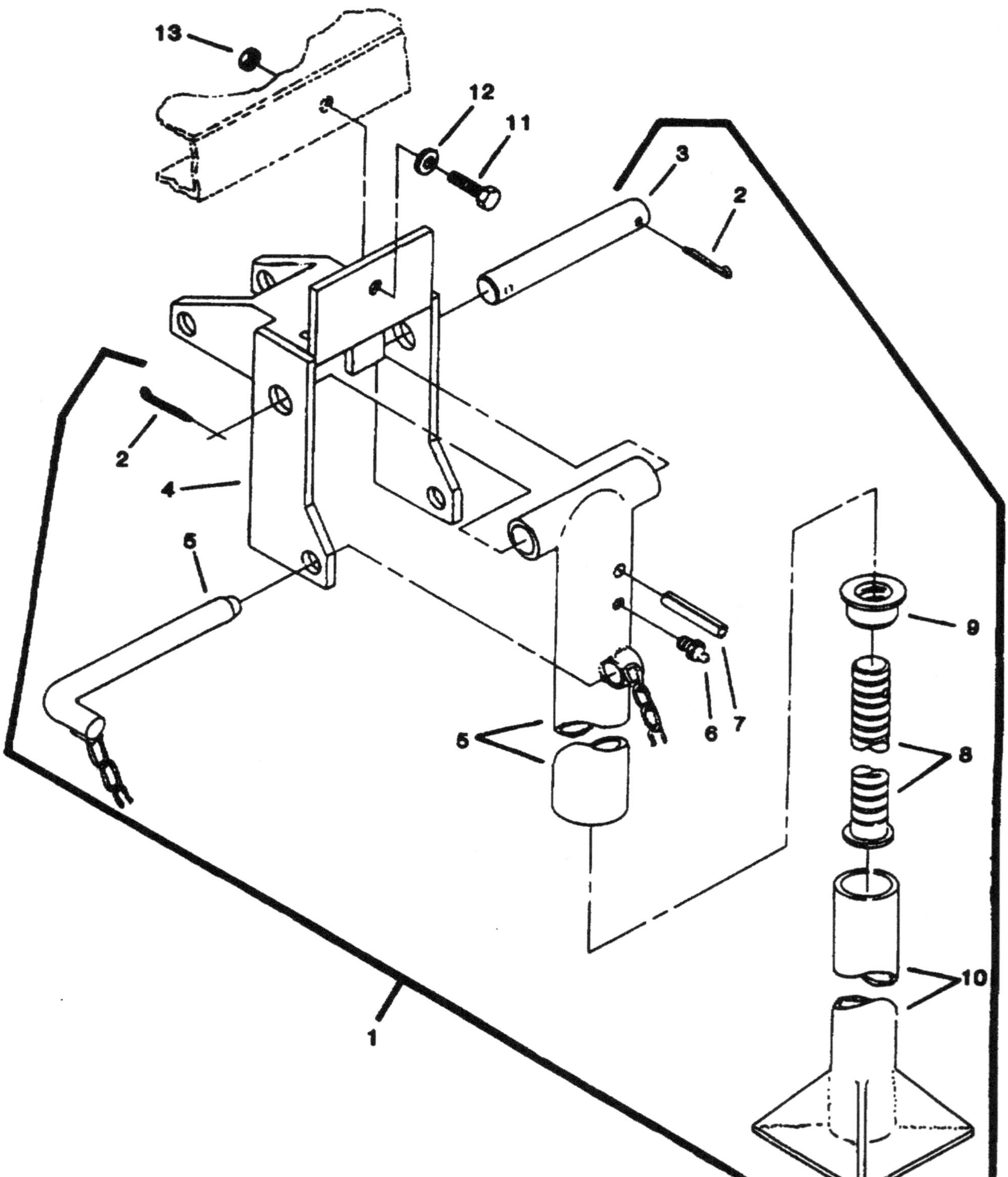

Figure C-11. Leg Prop Assembly.

(1) (2) (3) (4) (5) (6)

ITEM SMR PART
NO CODE FSCM NUMBER DESCRIPTION AND USABLE ON CODES(UOC) QTY

GROUP 06 TRAILER ASSEMBLY

FIG. C-11 LEG PROP ASSEMBLY

```
 1 PBOFZ 97403 13214E1206 .JACK,LEVELING-SUPPO .......................................................     1
 2 PAOZZ 96906 MS24665-353 ..PIN,COTTER SPLIT 1/8" X 1.0" LG.....................................     2
 3 XBFZZ    97403 13214E1209 ..PIN,STRAIGHT,HEADLE ......................................................     1
 4 XBOZZ 97403 13214E1207 ..BRACKET .....................................................................     1
 5 XBOZZ 97403 13214E1208 ..CHAIN,PIN RETAINING .......................................................     1
 6 PAOZZ 96906 MS15006-1 ..FITTING,LUBRICATION .......................................................     1
 7 PAOZZ 96906 MS16562-66 ..PIN,SPRING TUBULAR SLOTTED ..................................     1
 8 XDFZZ    97403 13214E1210 ..BOLT,MACHINE ..............................................................     1
 9 XDFZZ    97403 13214E1211 ..NUT,SLEEVE ..................................................................     1
10 XBFZZ         97403 13214E1212 ..SUPPORT BASE,LEG .............................................    1 11
PAOZZ 96906 MS90728-60 .SCREW,CAP,HEXAGON H 3/8-16 UNC-2A                                     31
                                        X 1.00" ...........................................................................
                                        UOC:EPH
11 PAOZZ 96906 MS90728-62 .SCREW,CAP,HEXAGON H 3/8-16 UNC-2A                                    3
                                        X 1.250" .........................................................................
                                        UOC:EPJ
12 PAOZZ 96906 MS27183-57 .WASHER,FLAT .406 NOM ID.......................................... 36
13 PAOZZ 96906 MS51922-17 .NUT,SELF-LOCKING,HE 3/8-16 UNC............................. 39
```

END OF FIGURE

C-41

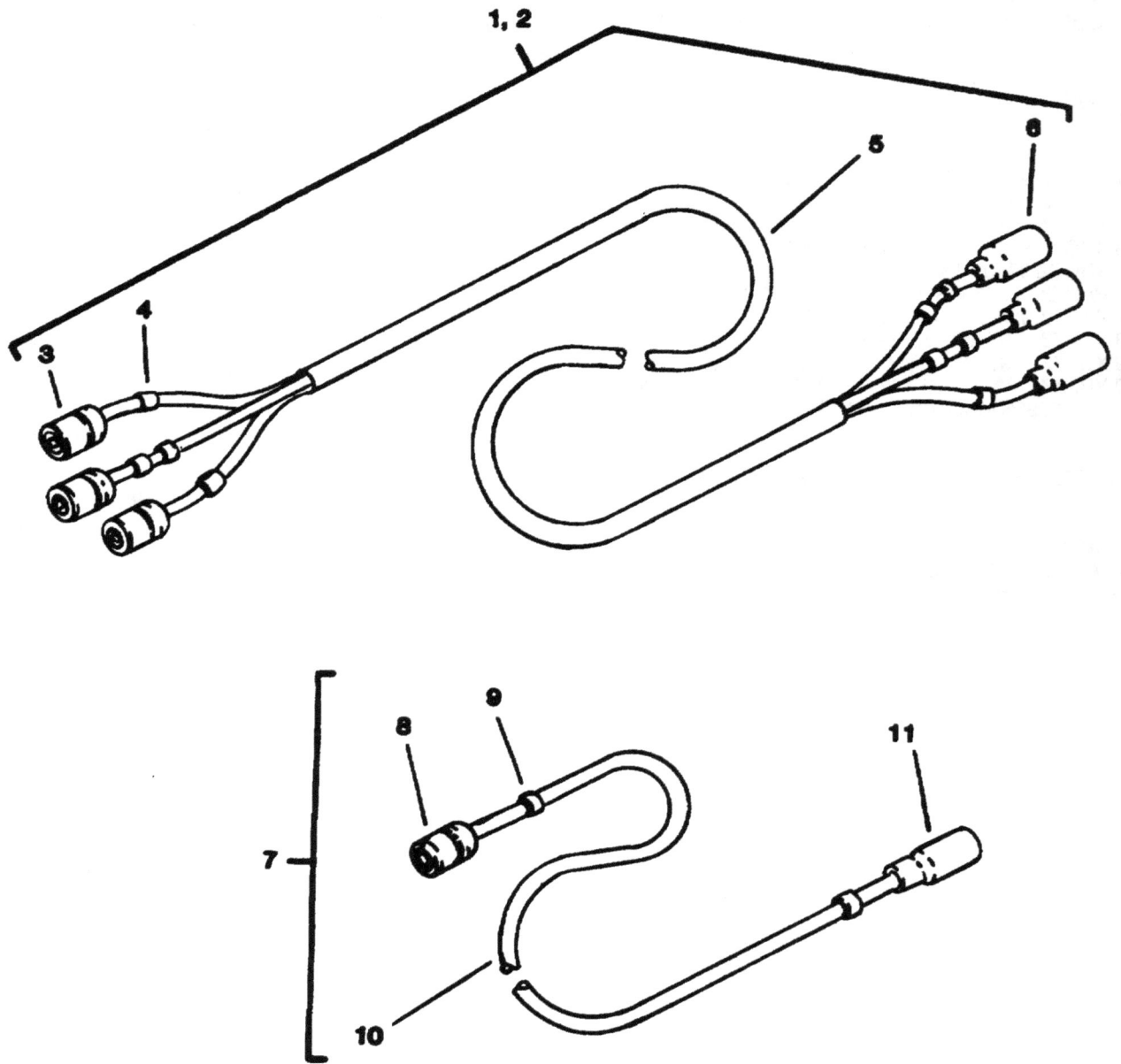

Figure C-12. Taillight Cable Assembly and Electrical Lead.

(1) (2) (3) (4) (5) (6)

ITEM SMR PART
 NO CODE FSCM NUMBER DESCRIPTION AND USABLE ON CODES(UOC) QTY

GROUP 06 TRAILER ASSEMBLY
FIG. C-12 TAIL LIGHT CABLE ASSEMBLY
AND ELECTRICAL LEAD

```
 1 MOOFF 97403 13216E7479-3 .CABLE ASSY ROADSIDE ....................................................   1
 2 MOOFF 97403 13216E7479-4 .CABLE ASSY CURBSIDE ....................................................   1
 3 PAOZZ 96906 MS27144-1 ..CONNECTOR,PLUG,ELEC ...............................................   6
 4 PAOZZ 81349 M43436/1-6 ..BAND,MARKER ...............................................................   9
 5 PAOZZ 81349 M13486/7-1 ..CABLE,SPECIAL PURPO ...................................................   1
 6 PAOZZ 96906 MS27142-2 ..CONNECTOR,PLUG,ELEC .................................................   6
 7 MOOFF 97403 13216E7476-1 .LEAD ELECTRICAL ..........................................................   2
 8 PAOZZ 96906 MS27144-1 ..CONNECTOR,PLUG,ELEC .................................................   2
 9 PAOZZ 81349 M43436/1-6 ..BAND,MARKER ...............................................................   2
10 PAOZZ 81349 M13486/1-5 ..WIRE,ELECTRICAL ...........................................................   2
11 PAOZZ 96906 MS27142-2 ..CONNECTOR,PLUG,ELEC....................................................   2
```

END OF FIGURE

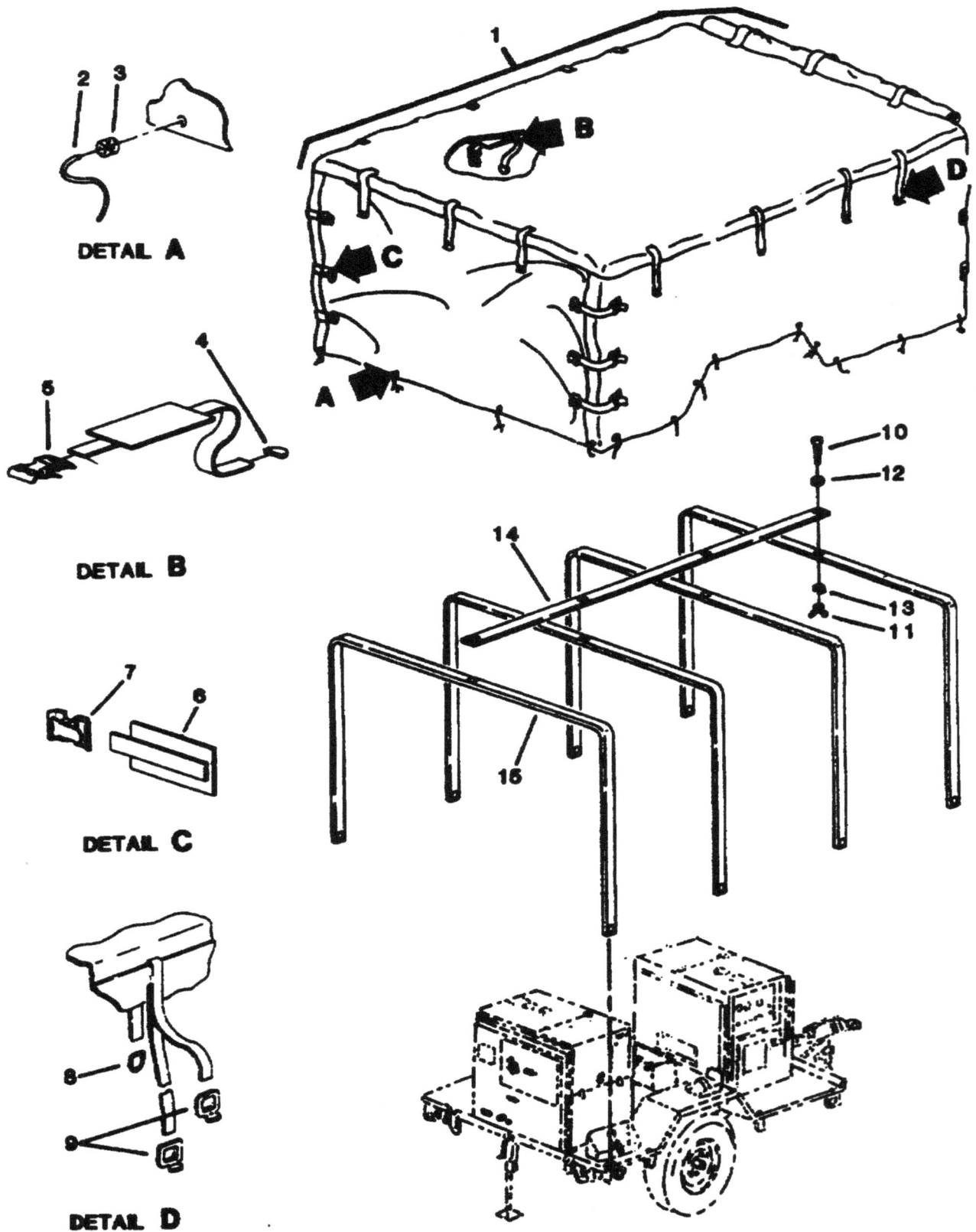

DETAIL A

DETAIL B

DETAIL C

DETAIL D

Figure C-13. Enclosure, AN/MJQ-33.

(1) (2) (3) (4) (5) (6)

ITEM SMR PART
NO CODE FSCM NUMBER DESCRIPTION AND USABLE ON CODES(UOC) QTY

GROUP 06 TRAILER ASSEMBLY

FIG. C-13 ENCLOSURE, AN/MJQ-33

(1)	(2)	(3)	(4) Description and Usable on Codes (UOC)	(5) QTY
1	XDOFF	97430 13214E1219	.FITTED COVER..	1 2
	PAOZZ	81348 T-R-605 TYS	..ROPE,FIBEROUS SISAL 3/8" NOM OD X 30' L..	V
3	PAOZZ	81349 MIL-G-16491 TYIC L3	..GROMMET,METALLIC TYPE 1, CLASS R.	27
4	PAOZZ	96906 MS51926-3	..CLIP,END,STRAP END STRAP	20
5	PAOZZ	96906 MS51929-2	..BUCKLE...	8 6
	MFOFF	97403 13214E1392	..CHAPE, ASSEMBLY REFER TO DWG 12 13214E1392	
7	PAFZZ	96906 MS51929-2	...BUCKLE...	12
8	PAFZZ	96906 MS51925-1	..RING-DEE MEDIUM STEEL, 5/8" 14 STRAP SIZE .... UOC:EPJ	
9	XDFZZ	97403 13226E0953	..HOOK, TEE	22
10	PAOZZ	96906 MS35206-268	.SCREW,NMACHINE *10-24 X 1.250"................................. STEEL .... UOC:EPJ	4
11	PAOZZ	96906 MS35425-68	.NUT,PLAIN,WING .190-24 UNC-2B................................... UOC:EPJ	4
12	PAOZZ	96906 MS27183-47	.WASHER,FLAT .... UOC:EPJ	4
13	PAOZZ	96906 MS35338-43	.WASHER,LOCK *10 REG STEEL .... UOC:EPJ	4
14	XBOZZ	97403 13221E4799	.SUPPORT,TARPAULIN.. UOC:EPJ	1
15	XBOZZ	97403 13214E1218-1	.BOW,TRAILER TARPAUL . ... UOC:EPJ	4

END OF FIGURE

(1) (2) (3) (4) (5) (6)

ITEM SMR PART
NO CODE FSCM NUMBER DESCRIPTION AND USABLE ON CODES(UOC) QTY

GROUP 09 BULK MATERIALS

FIG. BULK

```
 1 PAOZZ 81348 MMM-A-1617 ADHESIVE  TYPE 1 ....................................................     V 2
   PAOZZ 81348 MMM-A-1617 ADHESIVE  TYPE 1 ....................................................     V
                                            UOC:EPH
 3 PAFZZ       81349 MIL-C-20696 CLOTH COATED NYLON TY2 CL3 .....................................     V 4
   PAOZZ 81349 M23053/5-107-9 INSULATION SLEEVING HEAT SHRINK,                              V
                              WHITE ..........................................................
 5 PAOZZ 81349 M23053/5-107-5 INSULATION TUBING HEAT SHRINK ...................................     V
 6 PAOZZ 81349 MIL-R-6130 RUBBER,CELLULAR ......................................     V
 7 PAOZZ 84348 SN60WRMAP3 SOLDER LEAD-TIN 60/40 ......................................     V
 8 PAOZZ 96906 NS3367-1-9 STRAP,TIEDOWN,ELECT 6.30 STD CLR...........................     V
 9 PAOZZ 81348 V-T-295 TY1CL1 THREAD.NYLON SIZE FF,3 PLY,OD S1,                          V
                                    TY1,CL1 ...................................................
                                    UOC:EPH
10 PAFZZ       81348 V-T-295 TYICLI THREAD,NYLON SIZE FF,3 PLY,OD S1,                       V
                                    TYI1CL1.....................................................
11 PAOZZ 81348 V-T-295 THREAD,NYLON SIZE E 2 PLY TY4, OD                                V
                                    SHADE S1 ..................................................
12 PAOZZ 81349 MIL-W-530 TY2CL4 WEBBING TEXTILE...............................................     V
```

END OF FIGURE

C-46

Section III.

Special Tools List (Not Applicable)

C-47

CROSS-REFERENCE INDEXES
NATIONAL STOCK NUMBER INDEX

STOCK NUMBER FIG. ITEM STOCK NUMBER FIG. ITEM

STOCK NUMBER	FIG.	ITEM	STOCK NUMBER	FIG.	ITEM	STOCK NUMBER	FIG.	ITEM
5310-00-014-5850	5	4	5940-00-113-9826				3	5
	7	50	5940-00-143-4777	3	6			
	8	20	4	7		5940-00-021-3321		2 10 5940-00-143-
4794	3	7	5310-00-022-8834			2 13		46
5310-00-022-8847	2	20	6145-00-152-6499	12		10 5310-00-045-		
3296	2		8 5935-00-167-7775			12		3
	13			13	12 8			
5340-00-057-6956	6		6 4730-00-172-0049			11		6
	6		9 5310-00-184-8970			2 15 7 18 5310-00-		
184-8971	2 17 8 26 5310-00-187-2413		2 18 13			5		7
36 13	7		8	12 5310-00-059-9263		5		5
9905-00-202-3639	9 13 5305-00-068-0508		7			8		10
18 5305-00-068-0510	2 44 9905-00-205-2795		9			10		
	6 32			10	17			
	7		1 8040-00-221-3811 BULK		1			
	8		6 BULK		2			
	9		8 4210-00-223-4857		7	6		
	9 18		8		4			
	10 11 5305-00-225-3843		7 27					
	11		11 5310-00-225-6993			1		5
5305-00-068-0511	1i 11 5306-00-226-4825		5 10 5305-00-071-1324			6		3
5307-00-227-1741	2 24 5305-00-071-2067		1			3		7 42
5305-00-071-2506	10 16		8 18 5305-00-071-2510			6 28 5340-00-229-		
0340	5	2 5975-00-074-2072 BULK	8			6		12
5340-00-078-7029	6	77 45						
	7	15	8	22 8 24 5340-00-234-8422		5		6 13 4
5310-00-247-7186	2 14 5305-00-082-6721		6 20 5305-00-253-5614			2 41 5310-00-087-		
4652	6 29	7	28					
	7			4		8 9		
	8		3 5325-00-276-6059		8 27			
	9		7 6110-00-400-7592		2	1		
	9 16 6210-00-420-8628		2 28					
	10 5 2590-00-420-8929		11		1			
	10 9 5935-00-462-6603		12		6			
	11		13		12 11			
5310-00-088-1251	6 25 2590-00-473-6331		7 25					
	7 10		8		7 9 12 5305-00-543-4372			
	8		1 10		14 5310-00-543-4717			
	2 25 5940-00-105-6331		2 11		7	35 5940-00-113-8190		
	7 38		8		11 8 14 5310-00-582-5677			
	6 21							

CROSS-REFERENCE INDEXES
NATIONAL STOCK NUMBER INDEX

STOCK NUMBER FIG. ITEM STOCK NUMBER FIG. ITEM

STOCK NUMBER	FIG.	ITEM	STOCK NUMBER	FIG.	ITEM
5310-00-582-5965	6	18	5310-01-162-8569	2	22
	7	21	5330-01-162-8585	2	12
	7	33	5340-01-185-6239	6	26
1851	2	32	6145-00-705-6681	12	
1	1		5305-00-725-2317	9	
BULK	4		5320-00-753-3830	5	
3078	2	34	5310-00-809-4058	6	19
	7	22			
	7	34			
5310-00-809-5998	1	4			
5310-00-809-8546	2	7			
5305-00-821-3869	10	7			
5970-00-822-2775	3	9			
5315-00-838-4584	11	7			
5315-00-839-5822	11	2			
5310-00-877-5797	7	51			
	8	19			
5310-00-913-9776	7	41			
	8	17			
2590-00-932-7298	6	16			
	7	19			
5310-00-934-9758	2	9			
6210-00-941-6690	2	30			
5305-00-957-7086	5	3			
5340-00-975-2126	5	8			
5970-00-983-7985	BULK	5			
5310-00-984-3806	5	12			
5305-00-984-6214	2	6			
5305-00-984-6215	13	10			
5305-00-988-1724	9	9			
5305-00-988-1725	2	33			
5305-00-989-7434	2	27			
5305-00-993-1848	7	49			
	8	21			
5340-00-999-6277	7	31	5310-01-026-		
5824	2	19			
	7	37			
	8	13			
5340-01-026-8319	7	43			
	8	8			
5310-01-064-8787	6	17			
	7	20			
	7	32			
5310-01-106-1144	13	11			
5930-01-151-5442	2	37			
5930-01-160-0235	2	39			
6210-01-160-8026	2	29			

STOCK NUMBER	FIG.	ITEM
5310-00-584-7995	2	16
6210-01-230-	5	
6115-01-234-5966	5	
5970-00-740-2971	9	
5310-00-809-		

CROSS-REFERENCE INDEXES
NATIONAL STOCK NUMBER INDEX

CROSS-REFERENCE INDEXES
NATIONAL STOCK NUMBER INDEX
FSCM PART NUMBER STOCK NUMBER FIG. ITEM

96906 MS27142-2 5935-00-462-6603   12
5935-00-167-7775   12   3

96906 MS27183-10 5310-00-809-4058   6 19

96906 MS27183-11 5310-00-809-3078   2 34
96906 MS27183-18 5310-00-809-5998   1
96906 MS27183-42 5310-00-014-5850   5

96906 MS27183-47 13
52   6 27

96906 MS27183-56 5 11 96906 MS27183-57

96906 MS27183-8 5310-00-809-8546   2
96906 MS27969-4 5340-00-234-8422   5
96906 MS3367-1-9 5975-00-074-2072 BULK   8
96906 MS35206-267 5305-00-984-6214   2
96906 MS35206-268 5305-00-984-6215   13
96906 MS35206-280 5305-00-988-1724   9
96906 MS35206-281 5305-00-988-1725   2 33
96906 MS35207-263 5305-00-989-7434   2 27
96906 MS35207-265 5305-00-993-1848   7 49

96906 MS35333-108 5310-00-022-8834   2 13
96906 MS35333-110 5310-00-022-8847   2 20
96906 MS35335-91 5310-00-913-9776   7 41

96906 MS35338-101 5310-00-184-8970   2 15
96906 MS35338-103 5310-00-184-8971   2 17
96906 MS35338-43 5310-00-045-3296   2

96906 MS35338-44 5310-00-582-5965   6 18

96906 MS35387-1 9905-00-205-2795   9 10

96906 MS35387-2 9905-00-202-3639   9 13

11 96906 MS27144-1

12 8

7 22
7 34

4
4
7 50
8 20
12 96906 MS27183-

7 9
9 11
10 15
2 45
6 30
7 3
8 2
9 6
9 17
10 6
10 10
11 12
7
6

6
10
9

8 21

8 17

8
13 13

7 21
7 33

10 17

10 18

CROSS-REFERENCE INDEXES
NATIONAL STOCK NUMBER INDEX
FSCM PART NUMBER STOCK NUMBER FIG. ITEM

FSCM PART NUMBER STOCK NUMBER	FIG. ITEM	FIG. ITEM
96906 MS35425-28 5310-00-543-4717	2 25	
		7 35
		8 11
		11 96906 MS35425-
96906 MS35425-68 5310-01-106-1144	13	
70 5310-01-064-8787	6 17	
		7 20
		7 32
96906 MS35489-54 5325-00-276-6059	8 27	
96906 MS35649-202 5310-00-934-9758	2	9
96906 MS39347-2 5940-00-021-3321	2 10	
96906 MS51922-1 5310-00-088-1251	6 25	
		7 10
		9 12
		10 14
96906 MS51922-17 5310-00-087-4652	6 29	
		7 4
		8 3
		9 7
		9 16
		10 5
		10 9
		11 13
96906 MS51922-33 5310-00-225-6993	1	5
96906 MS51922-9 5310-00-984-3806	5 12	
96906 MS51925-1 13		8
96906 MS51926-3 5340-00-078-7029	6	7
		7 15
		8 24
		13 4
96906 MS51929-2 5340-00-057-6956	6	6
		6 9
		7 18
		8 26
		13 5
		13 7
96906 MS51939-3 5340-00-229-0340	5	2
		6 12 7 45
		8 22
96906 MS51957-81 5305-00-082-6721	6 20	
96906 MS51960-67 5305-00-071-1324	6	3
96906 NS53052-1 2590-00-473-6331	7 25	
		8 7
96906 MS90728-111 5305-00-071-2067	1	3
96906 MS90728-13 5305-00-071-2510	6 28	
96906 MS90728-3 5305-00-071-2506	10	16
96906 MS90728-32 5306-00-226-4825	5 10	
96906 MS90728-58 5305-00-543-4372	8	1
96906 MS90728-6 5305-00-068-0508	7	8
96906 MS90728-60 5305-00-068-0510	2 44	
		6 32

CROSS-REFERENCE INDEXES
NATIONAL STOCK NUMBER INDEX

FSCM PART NUMBER STOCK NUMBER FIG. ITEM

FSCM	PART NUMBER	STOCK NUMBER	FIG.	ITEM
96906	MS90728-60	5305-00-068-0510	7	1
				8 6
				9 8
				9 18
				10 ii
				11 11
96906	MS90728-62	5305-00-068-0511	1-1 11	
96906	MS90728-64	5305-00-725-2317	9	5
96906	MS90728-65	5305-00-821-3869	10	7
96906	MS90728-8	5305-00-225-3843	7 27	
96906	MS9319-208		7 17	
96906	MS9460-102	5		7
81349	M13486/1-5	6145-00-152-6499	12	10
81349	M13486/7-1	6145-00-705-6681	12	5
81349	M23053-15-105-5			4 8
81349	M23053/5-107-5	5970-00-983-7985 BULK	5	
81349	M23053/5-107-9			3 10
		5970-00-740-2971 BULK	4	
81349	M23053/5-110-9	5970-00-822-2775	3	9 81349 M43436/1-6
12	4			
				12 9
81349	MS086/2-10	4		5
80205	NASIS98-4Y 2 35			
58224	NE26G	2 31		
81348	QQW343C06B1B			3 11
				7 39
				8 15
81348	RR-C-271 TY2CL7			10 4
97403	SK-N-Q-002-TGN			7 44
84348	SN60WRMAP3 BULK		7	
81348	T-R-605 TYS 13			2
81348	V-T-295 BULK	11		
81348	V-T-295 TY1CL1 BULK			9
				BULK 10
97403	13205E4918 7 40			
				8 16
97403	13205E5078	5930-01-151-5442	2 37	
97403	13205E5079-3	6110-00-400-7592	2	1
97403	13205E5079-4			2 2
97403	13205E5120 6			2
97403	13205E5121 6 15			
97403	13205E5123 6 14			
97403	13205E5125	5340-01-185-6239	6 26	
97403	13205E5137-2			6 24
9'7403	13205E5143 7 26			
97403	13212E3553-1			7 30
97403	13212E3553-2			7 24
97403	13212E3560	6210-00-420-8628	2 28	
97403	13212E3567-1			4 1
97403	13212E3567-2			4 2
97403	13212E3567-3			4 3

CROSS-REFERENCE INDEXES
NATIONAL STOCK NUMBER INDEX
FSCM PART NUMBER STOCK NUMBER FIG. ITEM

FSCM	PART NUMBER	STOCK NUMBER	FIG.	ITEM
97403	13212E3567-4		4	4
97403	13212E3570-3		3	2
97403	13212E3570-4		3	4
97403	13212E3571-4		3	1
97403	13212E3571-5		3	3
97403	13212E3606	5940-00-105-6331	2	11
97403	13212E3610	5330-01-162-8585	2	12
97403	13212E3612	5340-01-026-8319	7	43
			8	8
97403	13212E3617	2590-00-932-7298	6	16
			7	19
97403	13214E1206	2590-00-420-8929	11	1
97403	13214E1207	11		4
97403	13214E1208	11		5
97403	13214E1209	11		3
97403	13214E1210	11		8
97403	13214E1211	11		9
97403	13214E1212	11		10
97403	13214E1214	5340-00-999-6277	7	31
97403	13214E1218-1	13	15	
97430	13214E1219	13		1
97403	13214E1223	530700-227-1741	2	24
			7	42
			8	18
97403	13214E1235	6210-00-223-4857	7	6
			8	4
97403	13214E1391	6210-00-900-9423	2	29
97403	13214E1392	13		6
97403	13216E7476-1	12		7
97403	13216E7479-3	12		1
97403	13216E7479-4	12		2
97403	13216E7504	6		8
97403	13216E75052		7	46
97403	13216E7600	2		3
97403	13216E7603	2	40	
97403	13216E7605	7	11	
97403	13216E7606-1		7	13
97403	13216E7607	7	12	
97403	13217E2062	7	14	
97403	13218E5091	6		4
			8	23
97403	13218E5139-1	5310-01-162-8569	2	22
97403	13218E5140-1		2	21
97403	13218E5149-2		2	23
97403	13218E5160	2	38	
97403	13219E9860	5930-01-160-0235	2	39
97403	13221E4799	13		14
97403	13221E7326	10		2
97403	13226E0953	13		9
97403	13226E5889-1		2	42
97403	13226E5889-2		2	43

CROSS-REFERENCE INDEXES
PART NUMBER INDEX

FSCM	PART NUMBER	STOCK NUMBER	FIG.	ITEM
97403	13226E7737	5	1	
97403	13228E6394-21		7	29
97403	13228E6394-22		8	10
97403	13228E9896	9	1	
97403	13228E9897-1		7	23
97403	13228E9897-2		7	2
97403	13228E9898	7	5	
97403	13228E9899	6	22	
97403	13228E9900-1	10	8	
97403	13228E9900-2		9	14
97403	13228E9901	9	3	
97403	13228E9902	6	1	
97403	13228E9903	9	4	
97403	13228E9904	9	2	
97403	13228E9905	9	15	
			10	12
97403	13228E9906	6	23	
97403	13228E9907	6	31	
97403	13229E2302	10	1	
97403	13229E2303-1		8	5
97403	13229E2303-2		7	7
97403	13229E2304	1	2	
72619	181-0937-0036210-00-941-6690		2	30
72619	181-8836-09-5536210-01-230-1851		2	32
19207	8747908-1	9	19	
			10	13

CROSS-REFERENCE INDEXES
FIGURE AND ITEM NUMBER INDEX

FIG. ITEM STOCK NUMBER FSCM PART NUMBER

BULK 1 8040-00-221-3811 81348 MMM-A-1617
BULK 2 8040-00-221-3811 81348 MMM-A-1617
BULK 3 81349 MIL-C-20696
BULK 4 5970-00-740-2971 81349 M23053/5-107-9
BULK 5 5970-00-983-7985 81349 M23053/5-107-5
BULK 6 81349 ML-R-6130
BULK 7 84348 SN60WRMAP3
BULK 8 5975-00-074-2072 96906 MS3367-1-9
BULK 9 81348 V-T-295 TY1CL1
BULK 10 81348 V-T-295 TY1CL1
BULK 11 81348 V-T-295
BULK 12 81349 MIL-W-530 TY2CL4 1 1 6115-01-234-5966 30554 MEP-
701A 1 297403 13229E2304 1 3 5305-00-071-2067 96906 MS90728-1 11 1 4 5310-00-
809-5998 96906 MS27183-18 1 5 5310-00-225-6993 96906 MS51922-33 2 1 6110-00-
400-7592 97403 13205E5079-3 2 2 97403 13205E5079-4 2 3
97403 13216E7600 2 4 81349 MIL-F-5591 2 5
96906 MS20427-4C6 2 6 5305-00-984-6214 96906 MS35206-267 2 7 5310-00-
809-8546 96906 MS27183-8 2 8 5310-00-045-3296 96906 MS35338-43 2 9
5310-00-934-9758 96906 M835649-202 2 10 5940-00-023-3321 96906 MS39347-2 2 11
5940-00-105-6331 97403 13212E3606 2 12 5330-01-162-8585 97403 13212E3610 2 13
5310-00-022-8834 96906 MS35333-108 2 14 5310-00-247-7186 96906 MS16203-37 2 15
5310-00-184-8970 96906 MS35338-101 2 16 5310-00-584-7995 96906 MS16203-27 2 17
5310-00-184-8971 96906 MS35338-103 2 18 5310-00-187-2413 88044 AN961-616T 2 19
5310-01-026-5824 96906 MS16203-39 2 20 5310-00-022-8847 96906 MS35333-110 2
21 97403 13218E5140-1 2 22 5310-01-162-8569 97403 13218E5139-1 2 23
97403 13218E5149-1 2 24 5307-00-227-1741 97403 13214E1223 2 25 5310-00-543-4717 96906 MS35425-28 2 26
81349 MIL-R-6130 2 27 5305-00-989-7434 96906 MS35207-263 2 28 6210-00-420-8628 97403 13212E3560 2 29
6210-00-900-9423 97403 13214E1391 2 30 6210-00-941-6690 72619 181-0937-003 2
31 58224 NE2G 2 32 6210-01-230-1851 72619 181-886-09-553 2 33 5305-00-
988-1725 96906 MS35206-281 2 34 5310-00-809-3078 96906 MS27183-11

CROSS-REFERENCE INDEXES
FIGURE AND ITEM NUMBER INDEX

FIG. ITEM STOCK NUMBER FSCM PART NUMBER

FIG.	ITEM	STOCK NUMBER	FSCM	PART NUMBER		
2	35		80205	NAS1598-4Y		
2	36		81349	MIL-R-6130		
2	37	5930-01-151-5442	97403	13205E5078		
2	38		97403	13218E5160		
2	39	5930-01-160-0235	97403	13219E9860		
2	40		97403	13216E7603		
2	41	5305-00-253-5614	96906	MS21318-20		
2	42		97403	13226E5889-1		
2	43		97403	13226E5889-2		
2	44	5305-00-068-0510	96906	MS90728-60		
2	45		96906	MS27183-57		
3	1				97403	13212E3571-4
3	2				97403	13212E3570-3
3	3				97403	13212E3571-5
3	4				97403	13212E3570-4
3	5	5940-00-113-9826	96906	MS25036-114		
3	6	5940-00-143-4777	96906	MS25036-157		
3	7	5940-00-143-4794	96906	MS25036-112		
3	8				81349	CO-04HDE
3	9	5970-00-822-2775	81349	M23053/5-110-9		
3	10		81349	M23053/5-107-9		
3	11		81348	QQW343C06B1B		
4	1				97403	13212E3567-1
4	2				97403	13212E3567-2
4	3				97403	13212E3567-3
4	4				97403	13212E3567-4
4	5				81349	M5086/2-10
4	6	5940-00-143-4794	96906	MS25036-112		
4	7	5940-00-143-4777	96906	MS25036-157		
4	8				81349	M23053-15-105-5
5	1				97403	13226E7737
5	2	5340-00-229-0340	96906	MS51939-3		
5	3	5305-00-957-7086	96906	MS24693-S273		
5	4	5310-00-014-5850	96906	MS27183-42		
5	5	5310-00-059-9263	96906	MS21046C3		
5	6	5340-00-234-8422	96906	MS27969-4		
5	7				96906	MS9460-102
5	8	5340-00-975-2126	96906	MS18015-1		
5	9	5320-00-753-3830	96906	MS20613-4P5		
5	10	5306-00-226-4825	96906	MS90728-32		
5	11		96906	MS27183-56		
5	12	5310-00-984-3806	96906	MS51922-9		
6	1				97403	13228E9902
6	2				97403	13205E5120
6	3	5305-00-071-1324	96906	MS51960-67		
6	4				97403	13218E5091
6	5				81349	MIL-W-4088
6	6	5340-00-057-6956	96906	MS51929-2		
6	7	5340-00-078-7029	96906	MS51926-3		
6	8				97403	13216E7504
6	9	5340-00-057-6956	96906	MS51929-2		

CROSS-REFERENCE INDEXES
FIGURE AND ITEM NUMBER INDEX
FIG. ITEM STOCK NUMBER FSCM PART NUMBER

FIG.	ITEM	STOCK NUMBER	FSCM	PART NUMBER		
6	10		81349	MIL-W-4088		
6	11		96906	MS24628-24		
6	12	5340-00-229-0340	96906	MS51939-3		
6	13		96906	MS24628-24		
6	14		97403	13205E5123		
6	15		97403	13205E5121		
6	16	2590-00-932-7298	97403	13212E3617		
6	17	5310-01-064-8787	96906	MS35425-70		
6	18	5310-00-582-5965	96906	MS35338-44		
6	19	5310-00-809-4058	96906	MS27183-10		
6	20	5305-00-082-6721	96906	MS51957-81		
6	21	5310-00-582-5677	96906	MS15795-810		
6	22		97403	13228E9899		
6	23		97403	13228E9906		
6	24		97403	13205E5137-2		
6	25	5310-00-088-1251	96906	MS51922-1		
6	26	5340-01-185-6239	97403	13205E5125		
6	27		96906	MS27183-52		
6	28	5305-00-071-2510	96906	MS90728-13		
6	29	5310-00-087-4652	96906	MS51922-17		
6	30		96906	MS27183-57		
6	31		97403	13228E9907		
6	32	5305-00-068-0510	96906	MS90728-60		
7	1	5305-00-068-0510	96906	MS90728-60		
7	2				97403	13228E9897-2
7	3				96906	MS27183-57
7	4	5310-00-087-4652	96906	MS51922-17		
7	5				97403	13228E9898
7	6	4210-00-223-4857	97403	13214E1235		
7	7				97403	13229E2303-2
7	8	5305-00-068-0508	96906	MS90728-6		
7	9				96906	MS27183-52
7	10	5310-00-088-1251	96906	MS51922-1		
7	11		97403	13216E7605		
7	12		97403	13216E7607		
7	13		97403	13216E7606-1		
7	14		97403	13217E2062		
7	15	5340-00-078-7029	96906	MS51926-3		
7	16		81349	MIL-W-530		
7	17		96906	MS9319-208		
7	18	5340-00-057-6956	96906	MS51929-2		
7	19	2590-00-932-7298	97403	13212E3617		
7	20	5310-01-064-8787	96906	MSS35425-70		
7	21	5310-00-582-5965	96906	MS35338-44		
7	22	5310-00-809-4058	96906	MS27183-10		
7	23		97403	13228E9897-1		
7	24		97403	13212E3553-2		
7	25	2590-00-473-6331	96906	MS53052-1		
7	26		97403	13205E5143		
7	27	5305-00-225-3843	96906	MS90728-8		
7	28	5305-00-253-5614	96906	MS21318-20		

CROSS-REFERENCE INDEXES
FIGURE AND ITEM NUMBER INDEX
FIG. ITEM STOCK NUMBER FSCM PART NUMBER

FIG.	ITEM	STOCK NUMBER	FSCM	PART NUMBER		
7	29		97403	13228E6394-21		
7	30		97403	13212E3553-1		
7	31	5340-00-999-6277	97403	13214E1214		
7	32	5310-01-064-8787	96906	MS35425-70		
7	33	5310-00-582-5965	96906	MS35338-44		
7	34	5310-00-809-4058	96906	MS27183-10		
7	35	5310-00-543-4717	96906	MS35425-28		
7	36	5310-00-187-2413	88044	AN961-616T		
7	37	5310-01-026-5824	96906	MS16203-39		
7	38	5940-00-113-8190	96906	MS25036-122		
7	39		81348	QQW343C06B1B		
7	40		97403	13205E4918		
7	41	5310-00-913-9776	96906	MS35335-91		
7	42	5307-00-227-1741	97403	13214E1223		
7	43	5340-01-026-8319	97403	13212E3612		
7	44		97403	SK-M-Q-002-TGM		
7	45	5340-00-229-0340	96906	MS51939-3		
7	46		97403	13216E7505-2		
7	47		81349	MIL-C-496		
7	48		81349	MIL-W-4088		
7	49	5305-00-993-1848	96906	MS35207-265		
7	50	5310-00-014-5850	96906	MS27183-42		
7	51	5310-00-877-5797	96906	MS21044-N3		
8	1	5305-00-543-4372	96906	MS90728-58		
8	2				96906	MS27183-57
8	3	5310-00-087-4652	96906	MS51922-17		
8	4	4210-00-223-4857	97403	13214E1235		
8	5				97403	13229E2303-1
8	6	5305-00-068-0510	96906	MS90728-60		
8	7	2590-00-473-6331	96906	MS53052-1		
8	8	5340-01-026-8319	97403	13212E3612		
8	9	5305-00-253-5614	96906	MS21318-20		
8	10		97403	13228E6394-22		
8	11	5310-00-543-4717	96906	MS35425-28		
8	12	5310-00-187-2413	88044	AN961-616T		
8	13	5310-01-026-5824	96906	MS16203-39		
8	14	5940-00-113-8190	96906	MS25036-122		
8	15		81348	QQW343CO6B1B		
8	16		97403	13205E4918		
8	17	5310-00-913-9776	96906	MS35335-91		
8	18	5307-00-227-1741	97403	13214E1223		
8	19	5310-00-877-5797	96906	MS21044-N3		
8	20	5310-00-014-5850	96906	MS27183-42		
8	21	5305-00-993-1848	96906	MS35207-265		
8	22	5340-00-229-0340	96906	MS51939-3		
8	23		97403	13218E5091		
8	24	5340-00-078-7029	96906	MS51926-3		
8	25		81349	MIL-W-4088		
8	26	5340-00-057-6956	96906	MS51929-2		
8	27	5325-00-276-6059	96906	MS35489-54		
9	1				97403	13228E9896

CROSS-REFERENCE INDEXES
FIGURE AND ITEM NUMBER INDEX
FIG. ITEM STOCK NUMBER FSCM PART NUMBER

FIG.	ITEM	STOCK NUMBER	FSCM	PART NUMBER		
9	2				97403	13228E9904
9	3				97403	13228E9901
9	4				97403	13228E9903
9	5	5305-00-725-2317	96906	MS90728-64		
9	6				96906	MS27183-57
9	7	5310-00-087-4652	96906	MS51922-17		
9	8	5305-00-068-0510	96906	MS90728-60		
9	9	5305-00-988-1724	96906	MS35206-280		
9	10	9905-00-205-2795	96906	MS35387-1		
9	11		96906	MS27183-52		
9	12	5310-00-088-1251	96906	MS51922-1		
9	13	9905-00-202-3639	96906	MS35387-2		
9	14		97403	13228E9900-2		
9	15		97403	13228E9905		
9	16	5310-00-087-4652	96906	MS51922-17		
9	17		96906	MS27183-57		
9	18	5305-00-068-0510	96906	MS90728-60		
9	19		19207	8747908-1		
10	1				97403	13229E2302
10	2				97403	13221E7326
10	3				96906	MS17990-C613
10	4				81348	RR-C-271 TY2CL7
10	5	5310-00-087-4652	96906	MS51922-17		
10	6				96906	MS27183-57
10	7	5305-00-821-3869	96906	MS90728-65		
10	8				97403	13228E9900-1
10	9	5310-00-087-4652	96906	MS51922-17		
10	10				96906	MS27183-57
10	11	5305-00-068-0510	96906	MS90728-60		
10	12				97403	13228E9905
10	13				19207	8747908-1
10	14	5310-00-088-1251	96906	MS51922-1		
10	15				96906	MS27183-52
10	16	5305-00-071-2506	96906	MS90728-3		
10	17	9905-00-205-2795	96906	MS35387-1		
10	18	9905-00-202-3639	96906	MS35387-2		
11	1	2590-00-420-8929	97403	13214E1206		
11	2	5315-00-839-5822	96906	MS24665-353		
11	3				97403	13214E1209
11	4				97403	13214E1207
11	5				97403	13214E1208
11	6	4730-00-172-0049	96906	MS15006-1		
11	7	5315-00-838-4584	96906	MS16562-66		
11	8				97403	13214E1210
11	9				97403	13214E1211
11	10				97403	13214E1212
11	11	5305-00-068-0510	96906	MS90728-60		
11	11	5305-00-068-0511	96906	MS90728-62		
11	12				96906	MS27183-57
11	13	5310-00-087-4652	96906	MS51922-17		
12	1				97403	13216E7479-3

CROSS-REFERENCE INDEXES
FIGURE AND ITEM NUMBER INDEX

FIG. ITEM STOCK NUMBER FSCM PART NUMBER

FIG.	ITEM	STOCK NUMBER	FSCM	PART NUMBER
12	3	5935-00-167-7775	96906	MS27144-1
12	4		81349	M43436/1-6
12	5	6145-00-705-6681	81349	M13486/7-1
12	6	5935-00-462-6603	96906	MS27142-2
12	7		97403	13216E7476-1
12	8	5935-00-167-7775	96906	MS27144-1
12	9		81349	M43436/1-6
12	10	6145-00-152-6499	81349	M13486/1-5
12	11	5935-00-462-6603	96906	MS27142-2
13	1		97430	13214E1219
13	2		81348	T-R-605 TYS
13	3		81349	MIL-G-16491 TY1C L3
13	4	5340-00-078-7029	96906	MS51926-3
13	5	5340-00-057-6956	96906	MS51929-2
13	6		97403	13214E1392
13	7	5340-00-057-6956	96906	MS51929-2
13	8		96906	MS51925-1
13	9		97403	13226E0953
13	10	5305-00-984-6215	96906	MS35206-268
13	11	5310-01-106-1144	96906	MS35425-68
13	12		96906	MS27183-47
13	13	5310-00-045-3296	96906	MS35338-43
13	14		97403	13221E4799
13	15		97403	13214E1218-1

C-61/(C-62 Blank)

APPENDIX D

COMPONENTS OF END ITEM LIST AND BASIC ISSUE ITEMS LIST

Section I. INTRODUCTION

D-1. Scope. This appendix lists components of end item and basic issue items for the power plants to help you inventory items required for safe and efficient operation.

D-2. General. The components of End Item and Basic Issue Items Lists are divided into the following sections:

a. Section II. Components of End Item. This listing is for informational purposes only, and is not authority to requisition replacements. These items are part of the end item, but are removed and separately packaged for transportation or shipment. As part of the end item, these items must be with the end item whenever it is issued or transferred between property accounts. Illustrations are furnished to assist you in identifying the items.

b. Section III. Basic Issue Items. These are the minimum essential items required to place the power plant in operation, to operate it, and to perform emergency repairs. Although shipped separately packaged, BII must be with the power plant during operation and whenever it is transferred between property accounts. The illustrations will assist you with hard-to-identify items. This manual is your authority to request/requisition replacement BII, based on TOE/MTOE authorization of the end item.

D-3. Explanation of Columns. The following provides an explanation of columns found in the tabular listings:

a. Column (1), Illustration Number (Illus No.). This column indicates the number assigned to the item.

b. Column (2), National Stock Number. Indicates the National stock number assigned to the item.

c. Column (3), Description. Indicates the federal item name and, if required, a minimum description to identify and locate the item. The last line for each item indicates the commercial and government entity (CAGE) (in parentheses) followed by the part number.

If item needed differed for different models of this equipment, the model would be shown under the "Usable on Code" heading in this column. The Usable on Code is not applicable for this equipment.

d. Column (4), Unit of Measure (U/M). Indicates the measure used in performing the actual operational/maintenance function. This measure is expressed by a two-character alphabetical abbreviation (eg, ea, in pr).

e. Column (5), Quantity Required (Qty Req'd). Indicates the quantity of the item authorized to be used with/on the equipment. The number 32 or 33 in parenthesis denotes the power plant.

D-1/(D-2 blank)

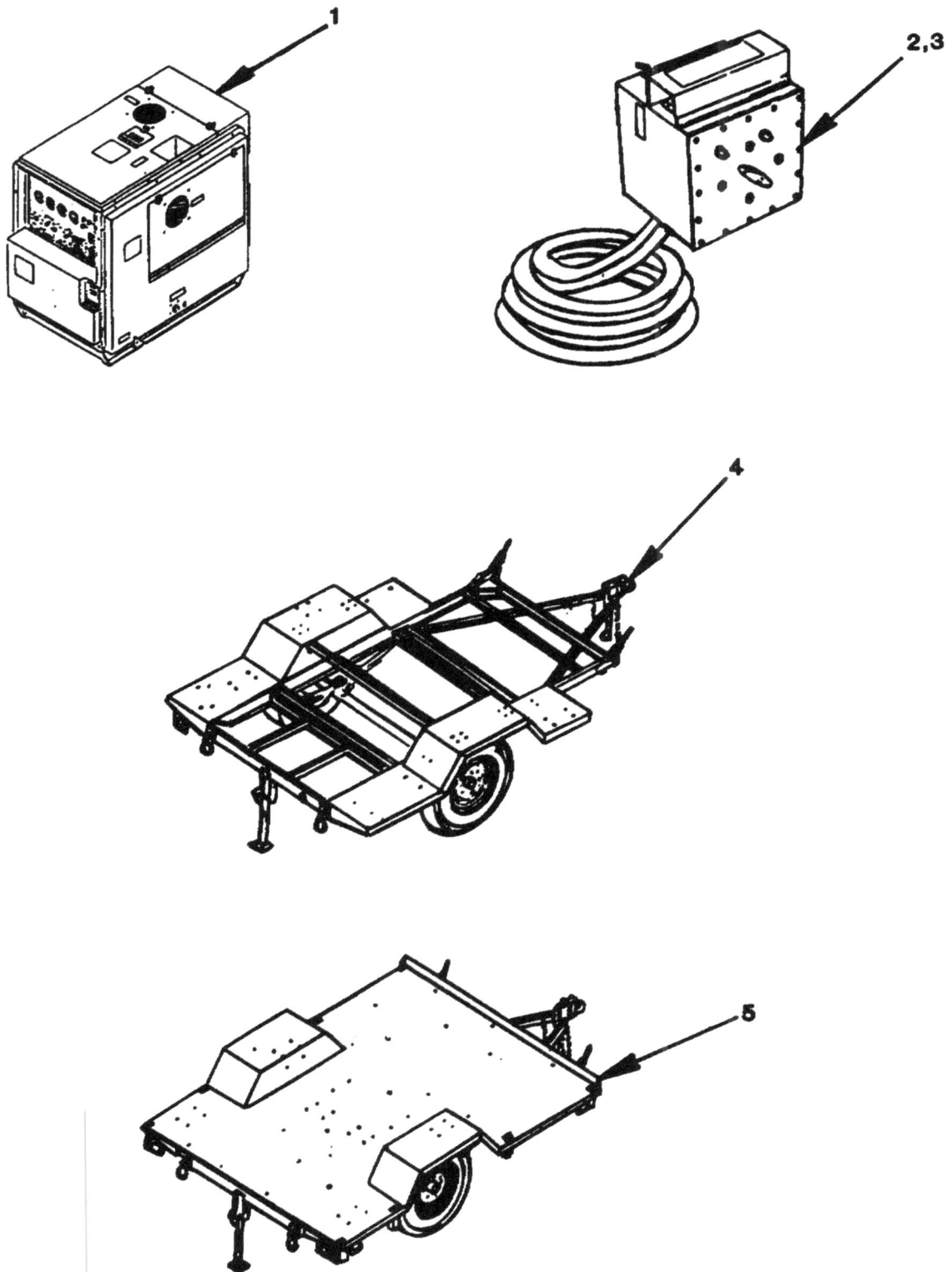

Figure D-1. Components of End Item (Sheet 1 of 2).

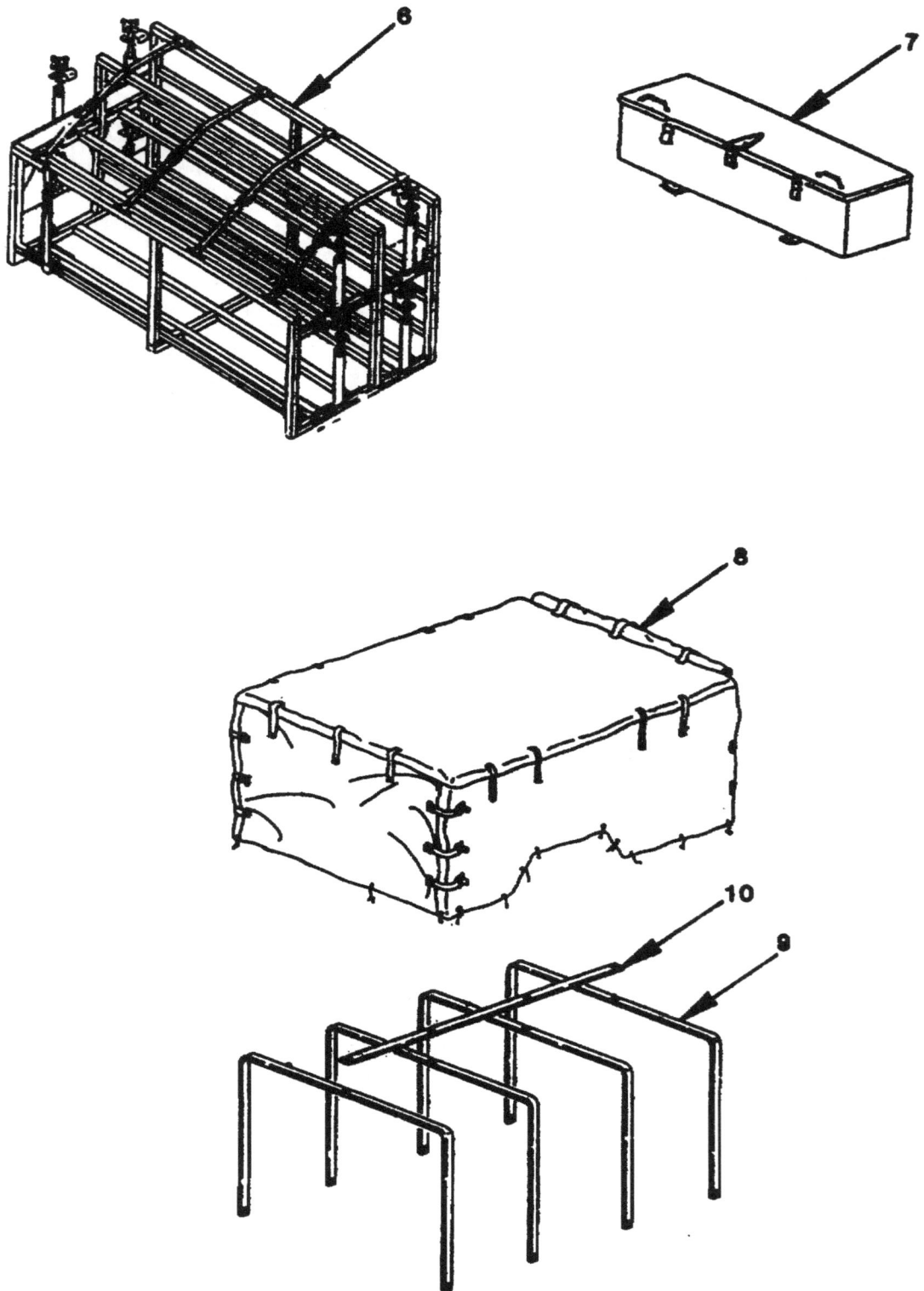

Figure D-1. Components of End Item (Sheet 2 of 2).

Section II. COMPONENTS OF END ITEM

(1) ILLUS. NO.	(2) NATIONAL STOCK NUMBER	(3) DESCRIPTION USABLE (CAGEC) AND PART NUMBER ON CODE	(4) U/M	(5) QTY REQD
1	6115-01-234-5966	Generator, Modified, 3kW, 60 Hz (30554) MEP-701A.	Ea	2 (32) 2 (33)
2		Box, Switch (974U3) 13205E5079-3	Ea	1 (32)
3		Box, Switch (97403) 13205E5079-4	Ea	1 (33)
4		Trailer, Modified, 3/4-ton (97403) 13228E9896	Ea	1 (32)
5		Trailer, Modified, 3/4-ton (97403) 13229E2302	Ea	1 (33)
6		Rack Assembly, Stowage (97403) 13228E9902	Ea	1 (32)
7	6115-01-230-0677	Box, Accessory (97403) 13226E7737	Ea	1 (33)
8	2540-00-926-0993	Tarpaulin, Fitted (97403) 13214E1219	Ea	1 (33)
9	2540-00-924-8478	Bow, Tarpaulin (97403) 13214E1218	Ea	4 (33)
10	2510-01-198-2885	Support, Tarpaulin (97403) 13221E4799	Ea	1 (33)

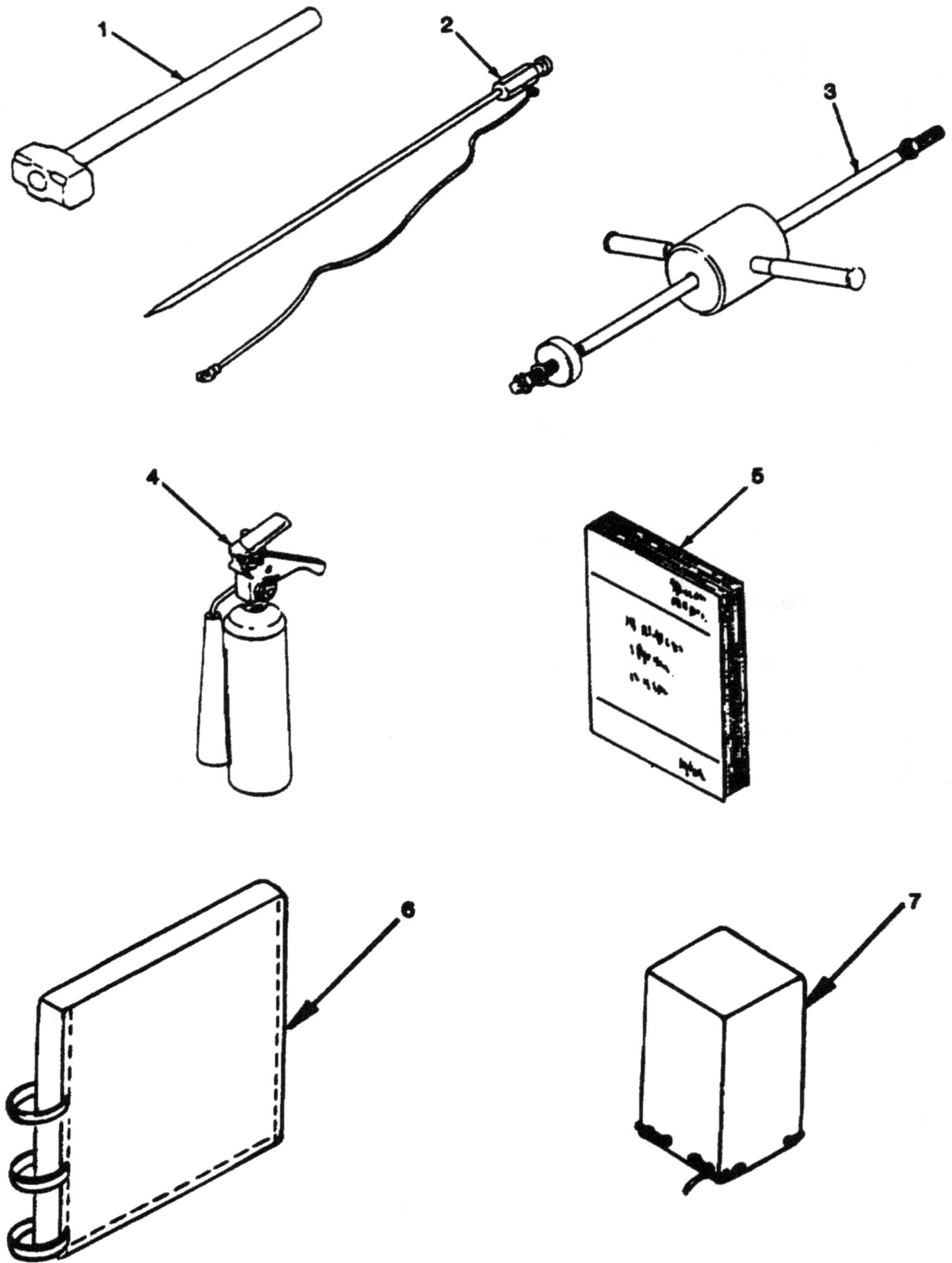

Figure D-2. Basic Issue Items (Sheet 1 of 2).

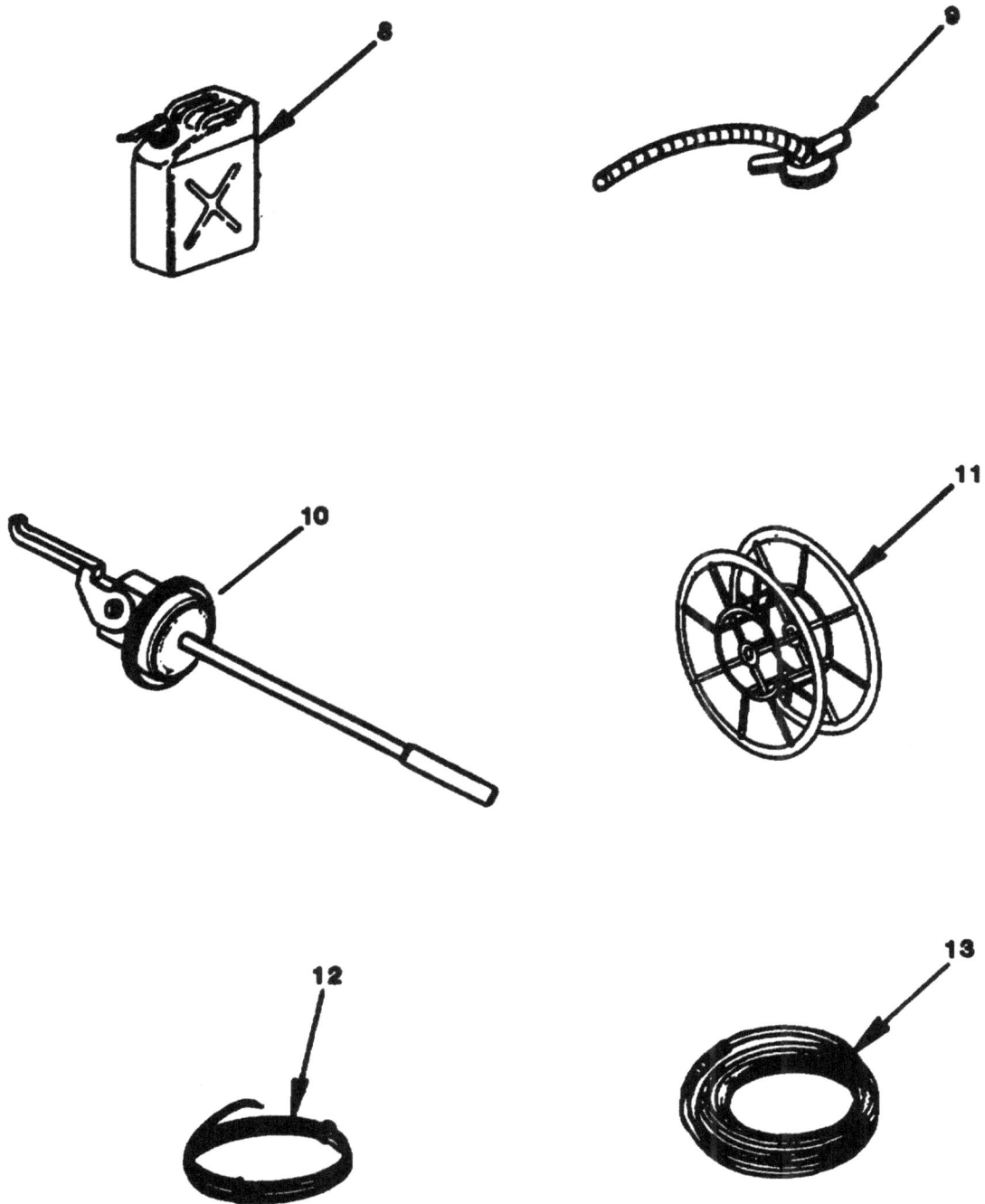

Figure D-2. Basic Issue Items (Sheet 2 of 2).

Section III. BASIC ISSUE ITEMS

(1) ILLUS. NO.	(2) NATIONAL STOCK NUMBER	(3) DESCRIPTION USABLE CAGE AND PART NUMBER ON CODE	(4) U/M	(5) QTY REQ'D
1	5120-00-2514489	Hammer Sledge, 8 lb(3.6kg) (81348) Type X, CL1	Ea	1(32) 1(33)
2	5975-00-878-3791	Rod, Ground, Driven, Three Sections 9 Ft (2.7 m) (81348) Type III Class B	Ea	2(32) 2(33)
3	5120-01-013-1676	Hammer, Slide (97403) 13226E7741	Ea	1(32) 1(33)
4	4210-00-555-8837	Extinguisher, Fire, Hand 4 lb (2.28 kg) (81348) MIL-E-52031	Ea	1(32) 1(33)
5		Manual, Technical TM5-6115-640-14&P	Ea	1(32) 1(33)
6		Cover, Antenna 13228E9908	Ea	2(32)
7		Cover, Switch Box 13228E9909	Ea	1(32)
8	7420-00-222-3088	Can, Fuel, Military 5 gal (18.9 L) (81349)	Ea	2(32) 4(33)
9	7420-00-177-6154	Spout, Can, Flexible (81349)	Ea	1(32) 1(33)
10	2910-00-066-1235	Adapter, Fuel Drum (97403)13211E7541	Ea	1(32) 1(33)
11	8130-00-656-1090	Reel, Cable (81349) RC-435/4	Ea	1(32)
12		Wire, Ground-1 (96")	Ea	2(32) 2(33)
		Wire Ground-2 (120")	Ea	1(33)
		Wire Ground-3 (12")	Ea	2(32) 2(33)

Change 2 D-8

Section III. BASIC ISSUE ITEMS - Continued

(1) ILLUS. NO.	(2) NATIONAL STOCK NUMBER	(3) DESCRIPTION USABLE CAGE AND PART NUMBER ON CODE	(4) U/M	(5) QTY REQ'D
		Wire, Ground-4 (28")	Ea	1(32)
		Wire, Ground-5 (12")	Ea	1(32) 1(33)
		Wire, Ground-6 (60")	Ea	1(33)
		Wire, Ground-7 (136")	Ea	1(32)
13	4720-00-021-3320	Hose, Fuel, Auxiliary 25 Ft (7.62 m)	Ea	2(32) 2(33)

D-9/(D-10 blank)

APPENDIX E

ADDITIONAL AUTHORIZATION

LIST
This appendix is not applicable.

APPENDIX F

EXPENDABLE SUPPLIES AND MATERIALS LIST

Section I. INTRODUCTION

F-1. SCOPE.

This appendix lists expendable supplies and materials you will need to operate and maintain the AN/MJQ-32 and AN/MJQ-33 Power Plants. These items are authorized to you be CTA 50-970, Expendable Items (Except Medical, Class V, Repair Parts, and Heraldic Items).

F-2. EXPLANATION OF COLUMNS .

a. Column (1) - Item Number. This number is assigned to the entry in the listing and is referenced in the narrative instructions to identify the material (e.g., "Use cleaning compound, item 5, App. F).

b. Column (2) - Level. This column identifies the lowest level of maintenance that requires the listed item.

(enter as applicable)

      C - Operator/Crew
      O - Organizational Maintenance
      F - Direct Support Maintenance
      H - General Support Maintenance

c. Column (3) - National Stock Number. This is the National stock number assigned to the item; use it to request or requisition the item.

d. Column (4) - Description. Indicates the Federal item name and, if required, a description to identify the item. The last line for each item indicates the Contractor and Government Entity (CAGE) in parentheses followed by the part number.

e. Column (5) - Unit of Measure (U/M). Indicates the measure used in performing the actual maintenance function. This measure is expressed by a two-character alphabetical abbreviation (e.g., ea, in, pr). If the unit of measure differs from the unit of issue, requisition the lowest unit of issue that will satisfy your requirements.

Section II. EXPENDABLE SUPPLIES AND MATERIALS LIST

(1) ITEM NUMBER	(2) LEVEL	(3) NATIONAL STOCK NUMBER	(4) DESCRIPTION	(5) U/M
1	C	6850-00-664-5685	Solvent, drycleaning (81348) PD-680	Qt
2	C	9150-00-186-6681	Oil, Lubricating (81349) OE/HDO-30	Qt
3	C	9150-00-265-9425	Oil, Lubricating (81349) OE/HDO-10	Qt
4	C	9150-00-402-4478	Oil, Lubricating (81349) OEA/APG-PD-1	Qt
5	O	9150-01-102-3658	Brake fluid, silicone (81349) BFS	Qt
6	O	9150-00-190-0904	Grease, automotive/artillery (81349) GAA	Lb
7	O		Tape, electrical	Ea
8	O	3439-00-273-2536	Solder	Rl

APPENDIX G

FABRICATION/ASSEMBLY OF PARTS

G-1. SCOPE.

This appendix includes complete instructions for making items authorized to be manufactured or fabricated at direct support maintenance.

G-2. GENERAL.

All bulk materials needed for manufacture of an item are listed by part number or specification number in a tabular list on the illustration.

G-3. MANUFACTURED ITEMS ILLUSTRATIONS .

See Figures G-1 thru G-5.

G-1

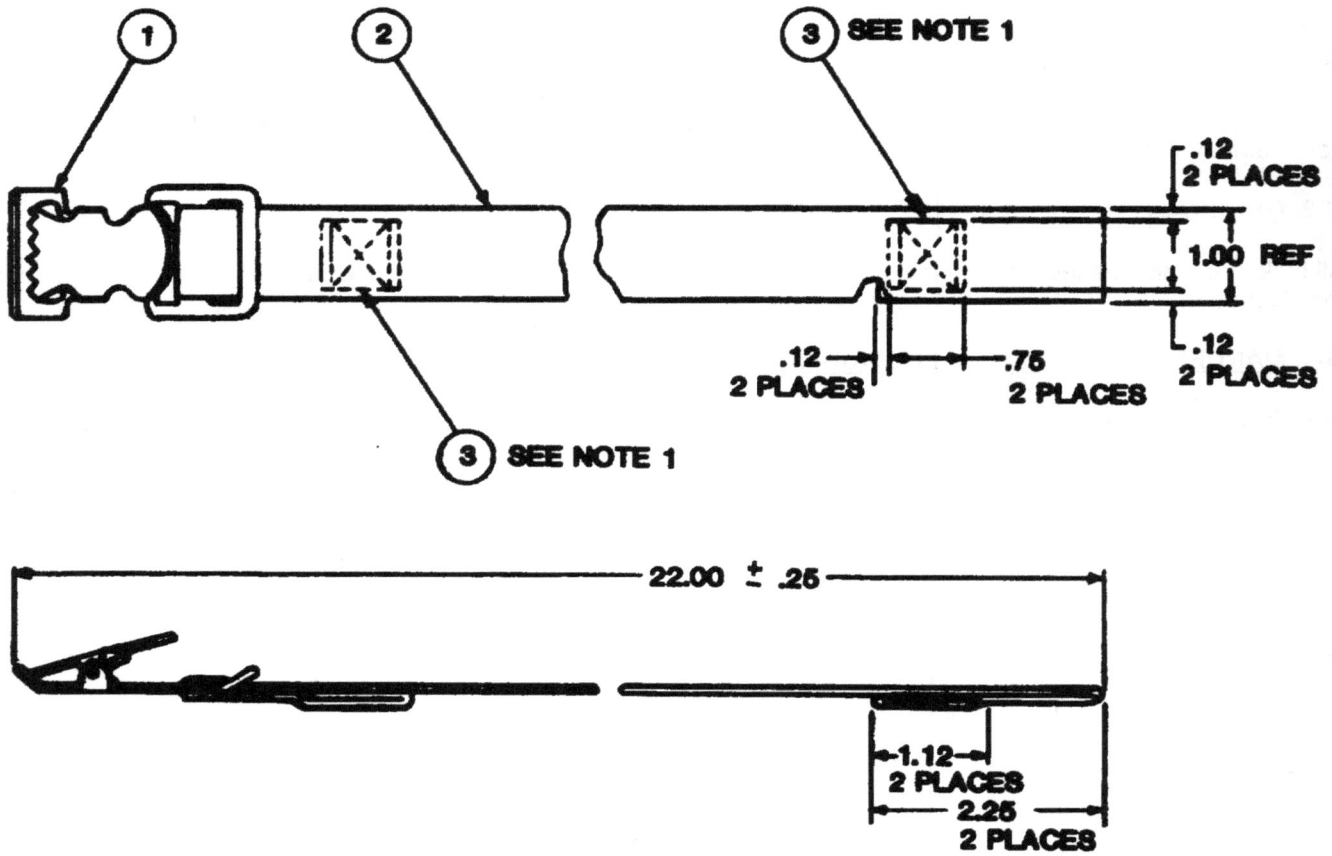

Figure G-1. Fabrication of Strapping on Stowage Rack for AN/MJQ-32 (Sheet 1 of 2).

13216E75U4 - Drawing Number

NOTE
All seams and stitching shall be in accordance with FED-STD-751. Class 300, Type 301, with a minimum of 10 stitches per inch. Bar tack or backstitch ends of all rows to prevent unraveling.

Find No.	Part or Identifying No.	Qty Req'd	Nomenclature or Description	Spec
1	MS51929-2	1	Buckle, spring action, CS, CADor ZN PL, size 1	
2	Type XVII	1	Webbing, textile, woven nylon, OD No. 7	MIL-W-4088
3	Type I, CL1	AR	Thread, nylon, size FF, OD No. 7	V-T-295

AR - As Required

Figure G-1. Fabrication of Strapping on Stowage Rack for AN/MJQ-32 (Sheet 2 of 2).

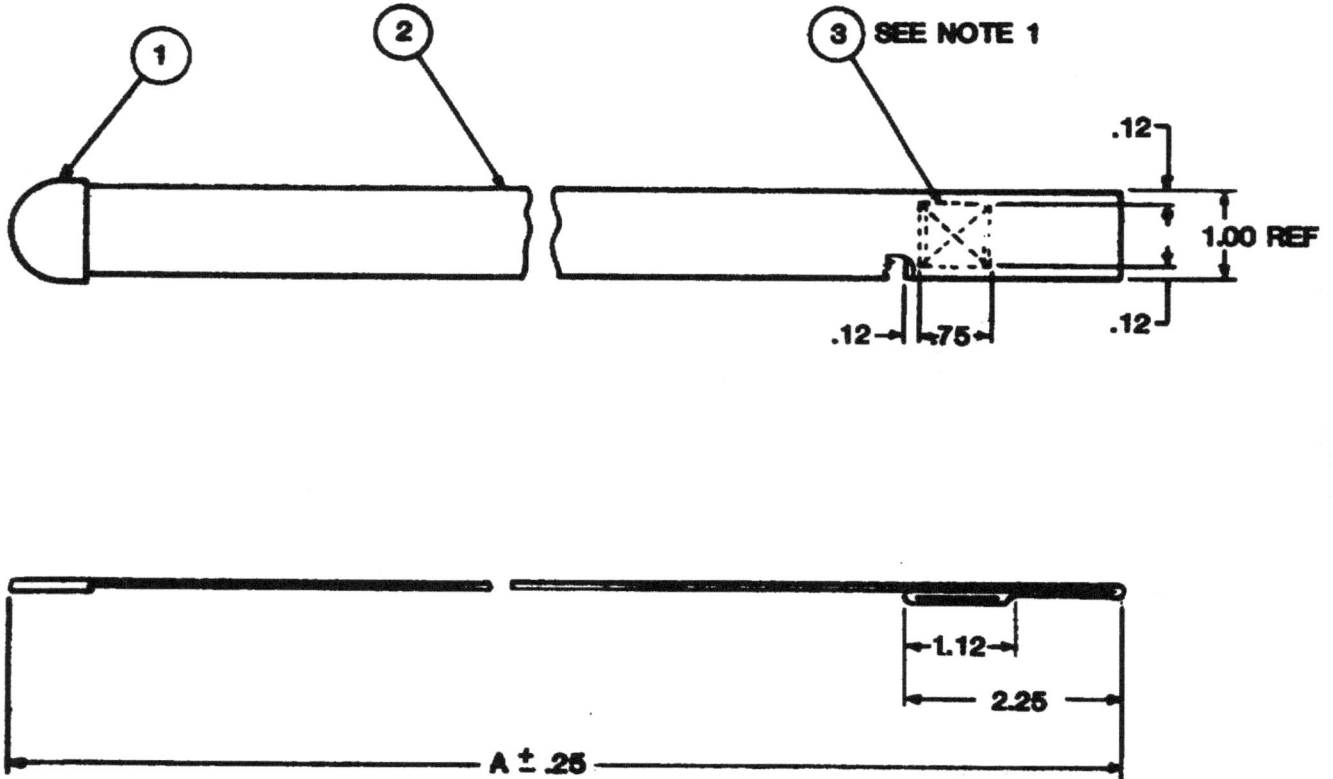

Figure G-2. Fabrication of Strapping on Trailer for AN/MJO-32 (Sheet 1 of 2).

13216E7505 - Drawing Number

NOTE
All seams and stitching shall be in accordance with FED-STD-751. Class 300, Type 301, with a minimum of 10 stitches per inch. Bar tack or backstitch ends of all rows to prevent unraveling.

Find No.	Part or Identifying No.	Qty Req'd	Nomenclature or Description	Spec
1	Type I, CL1	1	Clip, end, strap, size 1-inch	MIL-C-496
2	Type XVII	1	Webbing, textile, woven nylon, OD No. 7	MIL-W-4088
3	Type I, CL1	AR	Thread, nylon, size FF, OD No. 7	V-T-295

AR - As Required

Figure G-2. Fabrication of Strapping on Trailer for AN/MJQ-32 (Sheet 2 of 2).

G-5

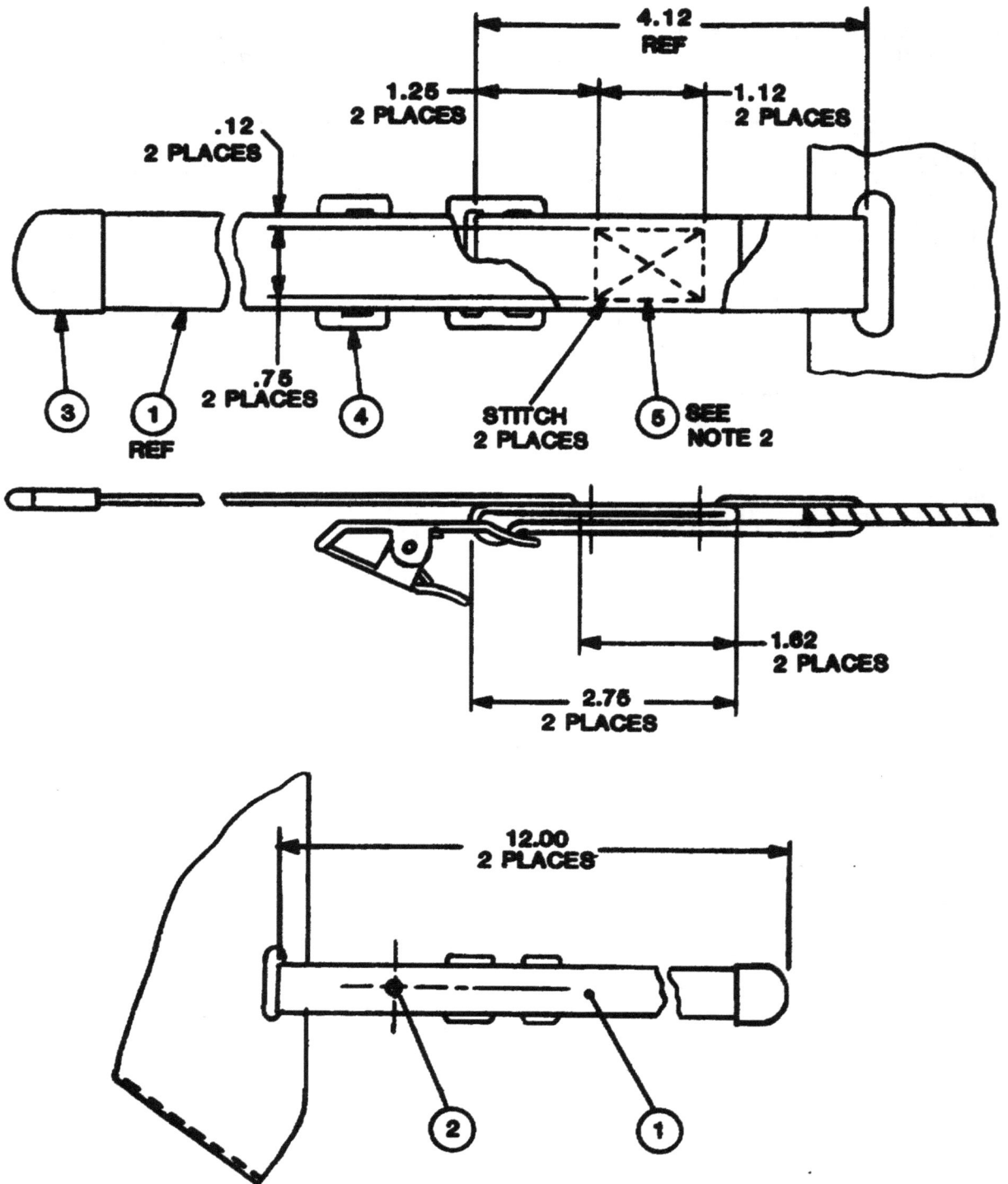

Figure G-3. Fabrication of Strapping on Cable-Reel Bracket for AN/MJQ-32 (Sheet 1 of 2).

13217E2062 - Drawing Number

NOTES:

1. Stitches shall be in accordance with FED-STD-751, Type 301, with 6 to 8 stitches per inch.

2. Find No. 1 thru 5 shall be free of paint.

Find No.	Part or Identifying No.	Qty Req'd	Nomenclature or Description	Spec
1	Type IIA	AR	Webbing and tape, textile,cotton, gen purpose, OD No. 7, 1" wide	MIL-W-530
2	MS9319-208	2	Rivet, solid, univ. ha, nickel-copper alloy, 3/16 dia. x .812 lg	
3	MS51926-3	2	Clip, end-strap, ball type, brass	
4	MS51929-2	2	Buckle, spring action, stl,cad, pl, 1.00 strap size	
5	Type I, Class I	AR	Thread and twine, mildew re-sis-tant	MIL-T-3530

AR - As Required

Figure G-3. Fabrication of Strapping on Cable-Reel Bracket for AN/MJQ-32 (Sheet 2 of 2).

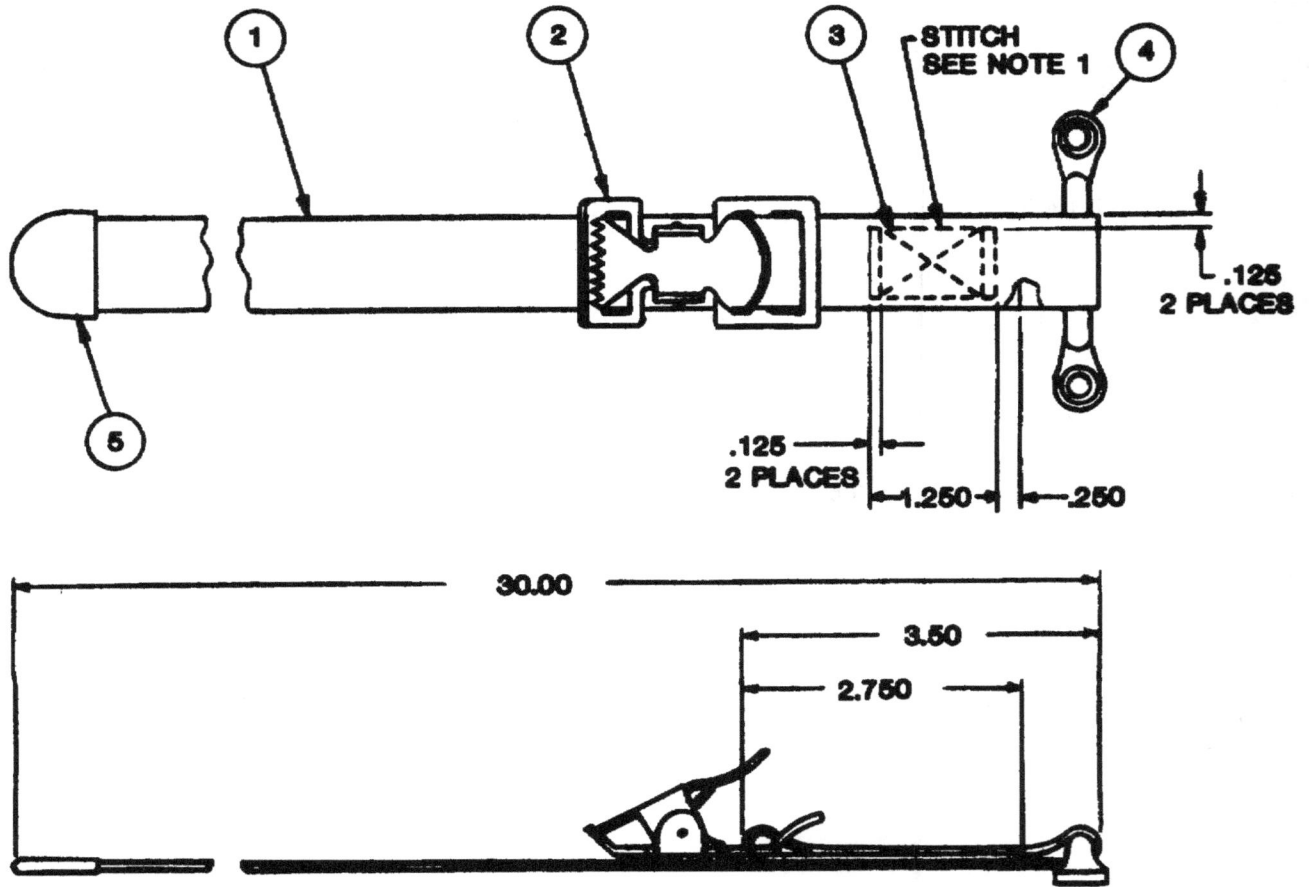

Figure G-4. Fabrication of Strapping on Bow Assembly Holddown for AN/MJQ-33 (Sheet 1 of 2).

13218E5091 - Drawing Number

NOTES:

1. All seams and stitching shall be in accordance with FED-STD-751.
    A. Stitches shall be type 301, 6-8 stitches per inch minimum.
    B. Bar tack or back stitch ends of all rows to prevent unraveling.
    C.   Stitching shall be type EFb-2.
1. Treat and paint find No. 2 and 5 in accordance with MIL-T-704, Type A, Color No. 24087.

Find No.	Part or Identifying No.	Qty Req'd	Nomenclature or Description	Spec
1	Type XVII	1	Webbing, 1 wide x 0.07 max thk, olive drab No. 7	MIL-W-4088
2	MS51929-2	1	Buckle, size 1	
3	Type I, CL1	AR	Thread, nylon, size FF, OD No. 7	V-T-295
4	MS51939-3	1	Loop, strap fastener	
5	MS51926-3	1	Clip, 1 strap fastener	

AR - As Required

Figure G-4. Fabrication of Strapping on Bow Assembly Holddown for AN/MJQ-33 (Sheet 2 of 2).

G-9

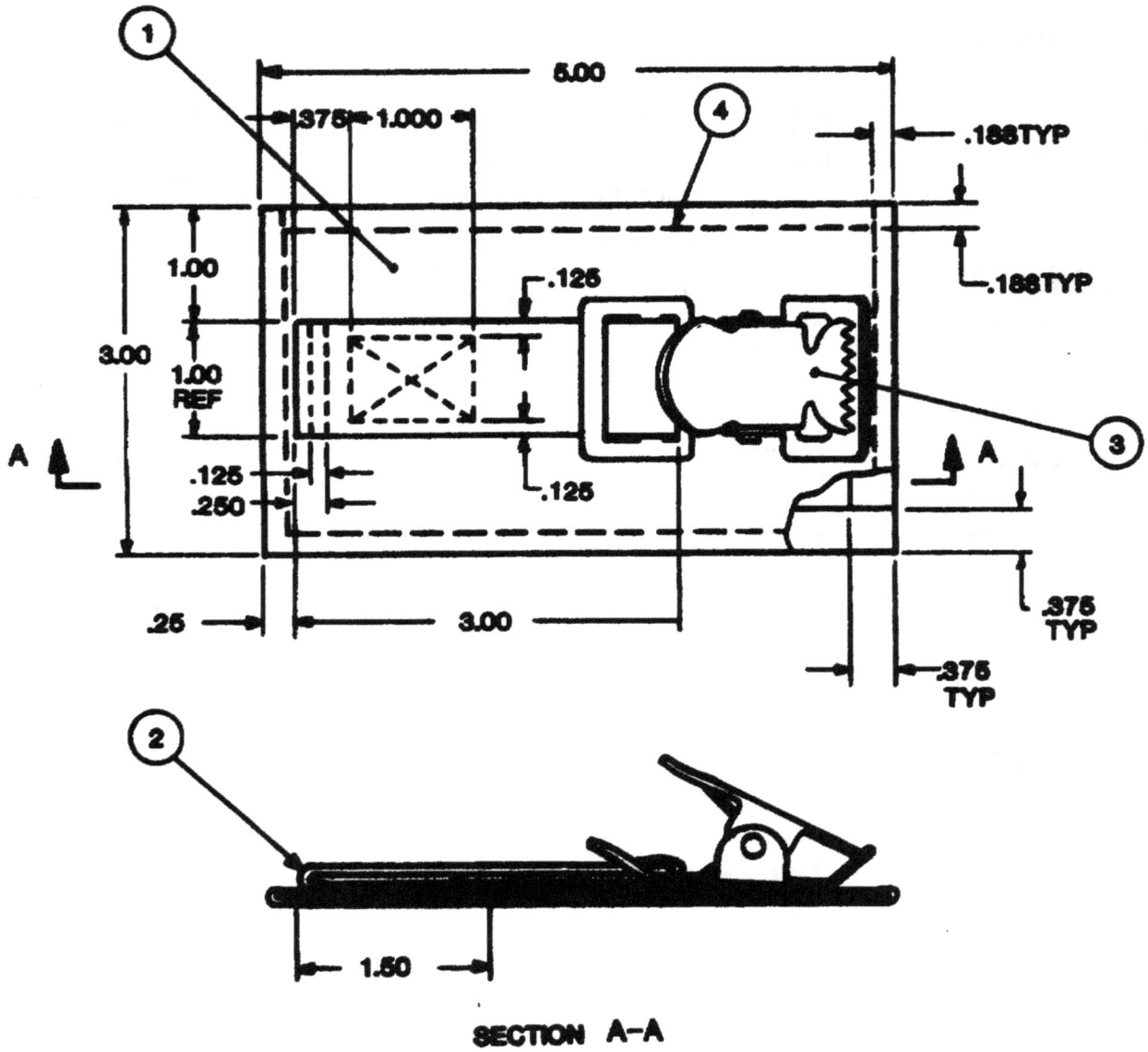

Figure G-5. Fabrication of Chafe Assembly on Tarpaulin for AN/MJQ-33 (Sheet 1 of 2)

13214E1392 - Drawing Number

NOTES:
1. For interpretation of: drawing, see DOD-STD-100.
2. All dotted lines indicate stitches.
3. Stitches shall be in accordance with FED-STD-751, Type 301, with 6 to 8 stitches per inch.
4. Ends of all stitch rows shall be back stitched.
5. Color shall be approximately OD color No. 34087 in accordance with FED-STD-595.

Find No.	Part or Identifying No.	Qty Req'd	Nomenclature or Description	Spec
1	Type II, Class 3	1	Cloth, coated nylon, waterproof,fire retardant, see note 5	MIL-C-20696
2	Type II, Class 4	AR	Webbing, textile, cotton, gen.purpose, med wt, dyed, waterrepellant, mildew	MIL-W-530
3	MS51929-2	1	Buckle, spring action, CS, CAD, PL, size 1	
4	Type I, Class 1	AR	Thread, nylon, TW soft multcord, low E long, size FF, 3ply OD shade S-1	VT-295

AR - As Required

Figure G-5. Fabrication of Chafe Assembly on Tarpaulin for AN/MJQ-33 (Sheet 2 of 2).

G-11/(G-12 blank)

INDEX

Subject Paragraph,

Index 1

INDEX

INDEX

Subject Paragraph,

Index 3/(Index 4 blank)

By Order of the Secretary of the Army:

CARL E. VUONO
General United States Army
Chief of Staff

Official:

WILLIAM J. MEEHAN, II
Brigadier General, United States Army
The Adjutant General

DISTRIBUTION:
  To be distributed in accordance with DA Form 12-25A, Operator, Unit, Direct Support and General Support Maintenance requirements for Generator Set, Diesel Engine Driven, Trailer Mounted

☆ **U.S. GOVERNMENT PRINTING OFFICE: 1996-406-421/60096**

SOMETHING WRONG WITH PUBLICATION

THEN...JOT DOWN THE DOPE ABOUT IT ON THIS FORM. CAREFULLY TEAR IT OUT, FOLD IT AND DROP IT IN THE MAIL.

FROM: (PRINT YOUR UNIT'S COMPLETE ADDRESS)

DATE SENT

PUBLICATION NUMBER	PUBLICATION DATE	PUBLICATION TITLE

BE EXACT   PIN-POINT WHERE IT IS

IN THIS SPACE, TELL WHAT IS WRONG AND WHAT SHOULD BE DONE ABOUT IT.

PAGE NO.	PARA-GRAPH	FIGURE NO.	TABLE NO.

TEAR ALONG PERFORATED LINE

PRINTED NAME, GRADE OR TITLE AND TELEPHONE NUMBER

SIGN HERE

DA FORM 1 JUL 79 2028-2

PREVIOUS EDITIONS ARE OBSOLETE.

P.S.--IF YOUR OUTFIT WANTS TO KNOW ABOUT YOUR RECOMMENDATION MAKE A CARBON COPY OF THIS AND GIVE IT TO YOUR HEADQUARTERS.

The Metric System and Equivalents

Linear Measure Liquid Measure

1 centiliter = 10 milliters = .34 fl. ounce
1 centimeter = 10 millimeters = .39 inch 1 deciliter = 10 centiliters = 3.38 fl. ounces
1 decimeter = 10 centimeters = 3.94 inches 1 liter = 10 deciliters = 33.81 fl. ounces
1 meter = 10 decimeters = 39.37 inches 1 dekaliter = 10 liters = 2.64 gallons
1 dekameter = 10 meters = 32.8 feet 1 hectoliter = 10 dekaliters = 26.42 gallons
1 hectometer = 10 dekameters = 328.08 feet 1 kiloliter = 10 hectoliters = 264.18 gallons
1 kilometer = 10 hectometers = 3,280.8 feet

Square Measure

Weights

1 sq. centimeter = 100 sq. millimeters = .155 sq. inch
1 centigram = 10 milligrams = .15 grain 1 sq. decimeter = 100 sq. centimeters = 15.5 sq. inches
1 decigram = 10 centigrams = 1.54 grains 1 sq. meter (centare) = 100 sq. decimeters = 10.76 sq. feet
1 gram = 10 decigram = .035 ounce 1 sq. dekameter (are) = 100 sq. meters = 1,076.4 sq. feet
1 decagram = 10 grams = .35 ounce 1 sq. hectometer (hectare) = 100 sq. dekameters = 2.47 acres
1 hectogram = 10 decagrams = 3.52 ounces 1 sq. kilometer = 100 sq. hectometers = .386 sq. mile
1 kilogram = 10 hectograms = 2.2 pounds
1 quintal = 100 kilograms = 220.46 pounds **Cubic Measure** 1 metric ton = 10 quintals = 1.1 short tons
1 cu. centimeter = 1000 cu. millimeters = .06 cu. inch
1 cu. decimeter = 1000 cu. centimeters = 61.02 cu. inches
1 cu. meter = 1000 cu. decimeters = 35.31 cu. feet

Approximate Conversion Factors

To change	To Multiply by To change	To Multiply by

inches centimeters 2.540 ounce-inches Newton-meters .007062
feet meters .305 centimeters inches .394
yards meters .914 meters feet 3.280
miles kilometers 1.609 meters   yards 1.094
square inches square centimeters 6.451 kilometers miles .621
square feet square meters .093 square centimeters square inches .155
square yards square meters .836 square meters square feet 10.764
square miles square kilometers 2.590 square meters square yards 1.196
acres square hectometers .405 square kilometers square miles .386
cubic feet cubic meters .028 square hectometers acres 2.471
cubic yards cubic meters .765 cubic meters cubic feet 35.315
fluid ounces milliliters 29,573 cubic meters cubic yards 1.308
pints liters .473 milliliters fluid ounces .034
quarts liters .946 liters pints 2.113
gallons liters 3.785 liters quarts 1.057
ounces grams 28.349 liters gallons .264
pounds kilograms .454 grams ounces .035
short tons metric tons .907 kilograms   pounds 2.205
pound-feet Newton-meters 1.356 metric tons short tons 1.102
pound-inches Newton-meters .11296

Temperature (Exact)

°F Fahrenheit 5/9 (after Celsius °C
temperature subtracting 32) temperature